PROLOG
FOREWORD

Die Erfolgsgeschichte der heutigen Dr. Ing. h.c. F. Porsche AG ist ohne das Lebenswerk von Ferry Porsche (1909–1998) nicht denkbar. Unter seiner Führung wurde aus dem 1931 gegründeten Konstruktionsbüro seines Vaters Ferdinand ein selbstständiger Automobilhersteller, der 1948 den ersten Sportwagen mit dem Namen Porsche baute. Mit dem Porsche 356 und später mit dem 911 realisierte Ferry Porsche unbeirrt seinen Traum vom »Fahren in seiner schönsten Form«.

Mit sicherem Gespür legte er die Grundlagen für die bis heute gültigen Markenwerte von Porsche. Als Geschäftsführer und Aufsichtsratsvorsitzender prägte er das Unternehmen über fünf Jahrzehnte hinweg. Vor allem aber ist es sein Verdienst als Visionär und Unternehmerpersönlichkeit, Porsche zu einem weltweit führenden Sportwagenhersteller entwickelt zu haben.

Dr. Ing. h.c. Ferdinand Anton Ernst »Ferry« Porsche an seinem 85. Geburtstag im Porsche Typ 356 »Nr.1«
Dr. Ing. h.c. Ferdinand Anton Ernst 'Ferry' Porsche in the Porsche Type 356 'No. 1' on his 85th birthday

The success that Dr. Ing. h.c. F. Porsche AG has achieved today would be unthinkable without the life's work that Ferry Porsche (1909–1998) devoted to the company. Under his management, the design office established in 1931 by his father Ferdinand Porsche grew into an independent automobile manufacturer. The first sports car bearing the Porsche name was built in 1948. With the Porsche 356 and the later 911, Ferry Porsche systematically transformed his dream of "driving in its finest form" into reality. With a sure hand, he shaped the fundamental principles that still govern Porsche's brand values today. As General Manager and later Chairman of the Supervisory Board, he gave the company its character for a period of fifty years. It is above all thanks to his efforts as a notable visionary and businessman that Porsche developed in to one of the world's leading sports car manufacturers.

Als Professor Dr. Ing. h.c. Ferdinand Anton Ernst »Ferry« Porsche am 27. März 1998 verstarb, war er einer der letzten großen Auto-Männer. Seine Lebensleistung für das Automobil brachte ihm einen Platz in der »European Automotive Hall of Fame« ein; sein Name wird heute in einem Atemzug genannt mit Größen wie Gottlieb Daimler, Carl Benz, Henry Ford oder Enzo Ferrari. Am 19. September 2009 wäre Ferry Porsche 100 Jahre alt geworden.

Die Porsche-Entwicklungsmannschaft des Typ 901 (1963)
The Porsche Type 901 development team (1963)

When Professor Dr. Ing. h.c. Ferdinand Anton Ernst Porsche died on March 27, 1998, the automotive world lost one of its last great men. His lifelong activity on behalf of the automobile earned him a place in the 'European Automotive Hall of Fame', and his name is mentioned in the same breath as other notable personalities such as Gottlieb Daimler, Carl Benz, Henry Ford and Enzo Ferrari. The centenary of Ferry Porsche's birth falls on September 19, 2009.

PROLOG | FOREWORD

PROLOG | FOREWORD

Empfang der Austro-Daimler-Rennmannschaft nach dem Klassensieg bei der 13. Targa Florio 1922. Hinter dem »Sascha«-Rennwagen mit der Startnummer 46: Ferry Porsche (3.v.l.) im Alter von 12 Jahren
At a reception for the Austro Daimler racing team after a class win in the 13th Targa Florio (1922). Behind the 'Sascha' racing car with start number 46: the 12-year-old Ferry Porsche (third from left)

»ICH WURDE MIT BENZIN AUFGEZOGEN UND BIN VOM AUTOMOBIL MEIN LEBEN LANG NICHT MEHR WEGGEKOMMEN.«

"I was weaned on gasoline and all my life I've never been able to drag myself away from the automobile."

»JEDER SOHN EINES GROSSEN MANNES HAT ES SCHWER.«
"Every son of a great man has a difficult time."

Ferry Porsche (rechts unten) mit seinem Vater Ferdinand Porsche 1950 im Porsche-Konstruktionsbüro in Stuttgart-Zuffenhausen
Ferry Porsche (below right) with his father Ferdinand Porsche at the Porsche Design Office in Stuttgart-Zuffenhausen (1950)

PROLOG | FOREWORD

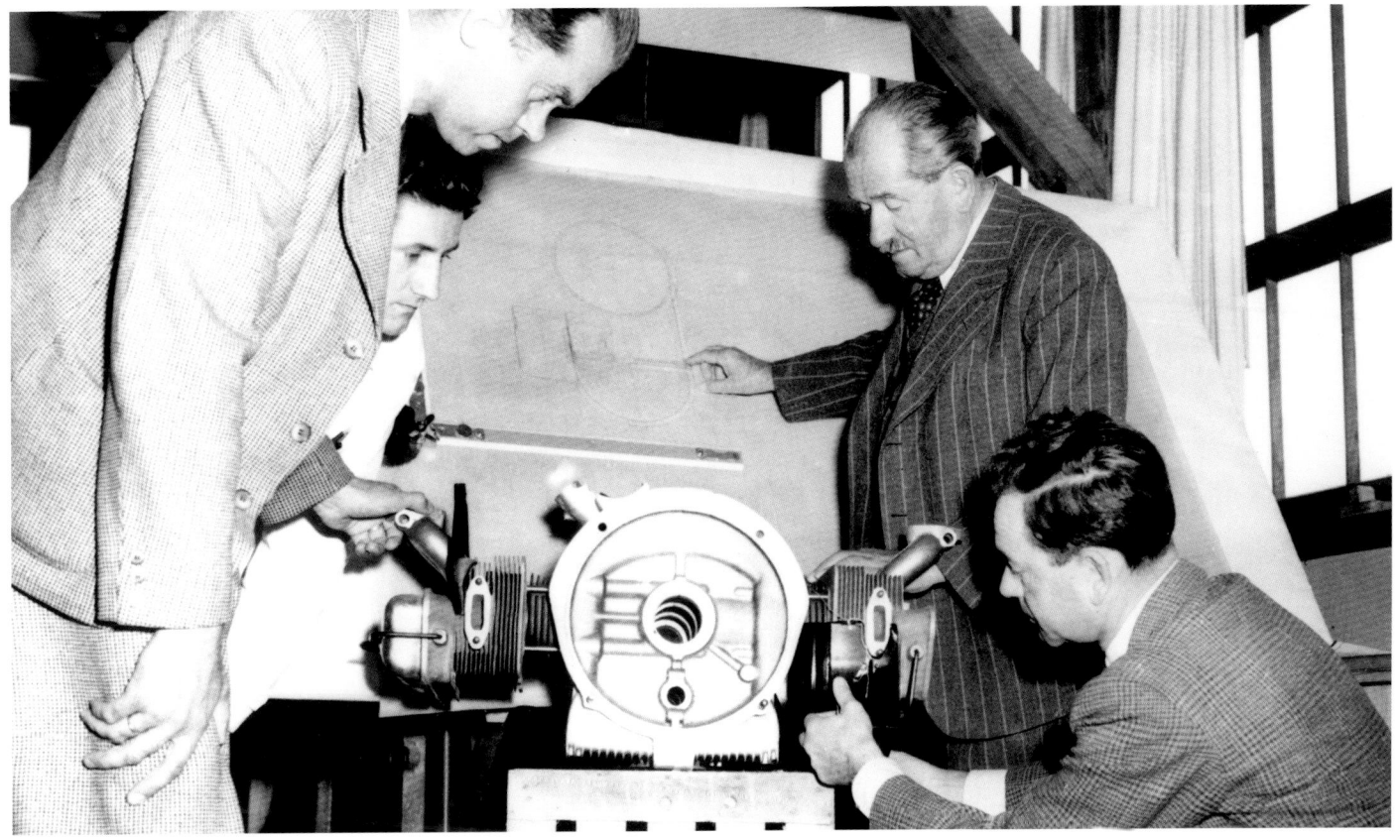

PROLOG | FOREWORD

Der erste Porsche-Sportwagen 1948 in Gmünd/Kärnten: Typ 356 »Nr.1«
The first Porsche sports car, the Type 356 'No.1' in Gmünd, Carinthia (1948)

»WIR HABEN EINFACH DAS AUTO GEBAUT, DAS WIR FÜR RICHTIG HIELTEN«
"We simply built the car we considered correct!"

»ICH HATTE DIE MARKTLÜCKE NOCH NICHT ENTDECKT, DA HATTE ICH SCHON SPASS, SO EIN AUTO ZU FAHREN.«
"Even before I discovered the market niche, I enjoyed driving a car like this."

Ferry Porsche in einem Porsche 356 A Cabriolet anlässlich der Enthüllung der Büste seines Vaters Ferdinand Porsche im VW-Werk in Wolfsburg am 3. September 1955
Ferry Porsche in a Porsche 356 A convertible at the ceremony on September 3, 1955 at which a bust of his father Ferdinand Porsche was unveiled at the VW plant in Wolfsburg

PROLOG | FOREWORD

24 Stunden-Rennen in Le Mans 1953: (V.r.n.l.) Rolf Wütherich, Richard von Frankenberg, Paul Frère, Willi Enz, Huschke von Hanstein, Bruno Trostmann, Hans Herrmann, Hubert Mimler, Ferry Porsche, Hugo Heiner, Wilhelm Hild, Helm Glöckler, Hans Klauser, Ernst Fuhrmann, (kniend) Herr Stubbe (Dunlop-Renndienst)

The 1953 Le Mans 24-hour race (from left to right): Rolf Wütherich, Richard von Frankenberg, Paul Frère, Willi Enz, Huschke von Hanstein, Bruno Trostmann, Hans Herrmann, Hubert Mimler, Ferry Porsche, Hugo Heiner, Wilhelm Hild, Helm Glöckler, Hans Klauser, Ernst Fuhrmann, (kneeling) and Mr. Stubbe (Dunlop Racing Service)

»**BEI PORSCHE NEHMEN WIR NICHT AN RENNEN TEIL, NUR UM UNSER SELBSTBEWUSSTSEIN ZU BEFRIEDIGEN, SONDERN UM UNS SELBST, UNSEREN VERSTAND UND DESHALB AUCH UNSERE AUTOS ZU VERBESSERN.**«

"At Porsche we don't enter races just to satisfy our self-esteem, but to improve ourselves, the state of our knowledge and the cars we build."

»ICH BIN DER MEINUNG, DASS DER SPORT-
WAGEN EIGENTLICH VORREITER FÜR NEUE
DINGE IST UND ES AUCH SCHON IN DER
VERGANGENHEIT WAR. ER WIRD ES AUCH
IN ZUKUNFT BLEIBEN.«
"In my opinion the sports car is the real forerunner of what's new.
It already was in the past, and it will remain that way in the future."

Übergabe eines Porsche 550 Spyder an seinen Besitzer Kurt Ahrens senior im Januar 1955 in Stuttgart-Zuffenhausen
Handing over a Porsche 550 Spyder to the owner, Kurt Ahrens Senior, in Stuttgart-Zuffenhausen (January 1955)

PROLOG | FOREWORD

PROLOG | FOREWORD

Ferry Porsche (rechts) 1956 mit seinem Chefkonstrukteur Karl Rabe (links) und dem Leiter der Karosserieentwicklung Erwin Komenda (Mitte)
Ferry Porsche (right) in 1956, with his Chief Designer Karl Rabe (left) and the Body Development Manager Erwin Komenda (center)

»DAS PERFEKTE AUTO GIBT ES NICHT. ABER WIR ALS INGENIEURE UND DESIGNER MÜSSEN ALLES TUN, UM DIESEM IDEAL IMMER NÄHER ZU KOMMEN.«

"There's no such thing as the perfect car. But as engineers and designers, we must do everything we can to get closer to this ideal all the time."

»DIE LANGE, GERADEZU UNGEWÖHNLICHE LEBENSDAUER DIESES MODELLS MACHT MICH DOCH STOLZ DARAUF, AM ENDE MIT MEINER MEINUNG VOM 911 RECHT BEHALTEN ZU HABEN.«

"Now that the 911 has been on the market for such an exceptionally long time, I'm proud that my opinion about it has proved to be right."

Ferry Porsche 1968 mit seinem Sohn Ferdinand Alexander an einem Porsche 911 im Fabrikhof Werk II in Stuttgart-Zuffenhausen
Ferry Porsche in 1968, seen with his son Ferdinand Alexander and a Porsche 911 in the yard of #2 plant in Stuttgart-Zuffenhausen

24 Stunden-Rennen von Le Mans 1970:
Ferry Porsche beobachtet in der Renn-Werkstatt in Teloché die Vorbereitungsarbeiten an einem Porsche 917 KH

The 1970 Le Mans 24-hour race: at the racing department's workshop in Teloché, Ferry Porsche watches the preparation of a Porsche 917 KH

»ICH VERSUCHE, MICH IM UMGANG MIT MEINEN MITARBEITERN STETS VON LOYALITÄT UND MENSCHLICHKEIT LEITEN ZU LASSEN.«

"In my dealings with colleagues I always try to be guided by principles of loyalty and humanity."

»ÜBER DIE REZEPTE, ERFOLGREICH ZU SEIN, GIBT ES UNZÄHLIGE BÜCHER. IN IHNEN STEHEN ALLE MÖGLICHEN GUTEN TIPS. ICH HABE KEINES DIESER BÜCHER GELESEN.«
"So many books have been written containing recipes for success They're full of useful hints. I haven't read any of them!"

Ferry Porsche an seinem 85. Geburtstag am 19. September 1994 auf dem Schüttgut in Zell am See
Ferry Porsche's 85th birthday on September 19, 1994 at the 'Schüttgut' in Zell am See

 **2** **PROLOG | FERRY PORSCHE – 100 JAHRE**
FOREWORD | FERRY PORSCHE – 100 YEARS

 **28** **EINE KINDHEIT UNTER AUTOMOBILEN**
CHILDHOOD AMONG AUTOMOBILES

 **40** **JUGENDZEIT IN STUTTGART**
YOUTH IN STUTTGART

PROLOG | FOREWORD

54 LEHRJAHRE IM KONSTRUKTIONSBÜRO
EARLY YEARS IN THE DESIGN OFFICE

94 DER ERSTE PORSCHE-SPORTWAGEN
THE FIRST PORSCHE SPORTS CAR

130 DER UNTERNEHMER FERRY PORSCHE
FERRY PORSCHE – THE BUSINESSMAN

182 LEBENSLAUF
PERSONAL CHRONICLE

EINE KINDHEIT UNTER AUTOMOBILEN
CHILDHOOD AMONG AUTOMOBILES

Ferry Porsche mit seiner
Schwester Louise in einem
Austro-Daimler Prinz-
Heinrich-Wagen (1912)
*Ferry Porsche with his
sister Louise. In 1912,
in an Austro-Daimler
'Prince Henry' car*

EINE KINDHEIT UNTER AUTOMOBILEN
CHILDHOOD AMONG AUTOMOBILES

Als Sohn des Automobilkonstrukteurs Ferdinand Porsche und seiner Frau Aloisia wurde Ferry Porsche am 19. September 1909 in Wiener Neustadt, Österreich, geboren. Dass das Automobil sein Leben prägen sollte, ließ sich bereits am Tag seiner Geburt erahnen: Als er das Licht der Welt erblickte, saß sein Vater Ferdinand gerade am Steuer eines von ihm konstruierten Austro-Daimler-Rennwagens und erzielte einen Klassensieg beim Semmering-Bergrennen. Getauft wurde der Porsche-Stammhalter auf den Namen Ferdinand Anton Ernst, doch seinen lebenslangen Rufnamen erhielt er von seinem Kindermädchen: Sie nannte ihn als Erste »Ferry«. Zusammen mit seiner fünf Jahre älteren Schwester Louise wuchs Ferry Porsche in einem behüteten Elternhaus auf, in dem das Thema Automobil den Lebensmittelpunkt bildete. Als Chefkonstrukteur

Bruder und Schwester im Jahr 1910
Brother and sister, 1910

Ferry Porsche was born on September 19, 1909 in the town of Wiener Neustadt, Austria, the son of automobile designer Ferdinand Porsche and his wife Aloisia. All the evidence suggests that the automobile was destined from the start to influence his life: as his wife was giving birth to their son, father Ferdinand was at the wheel of an Austro-Daimler competition car of his own design on the Semmering hillclimb, where he scored a class win. Porsche's son and heir was given the names Ferdinand Anton Ernst, but acquired the familiar form of his first name that was to accompany him throughout his life from his nursemaid, who was the first person to call him "Ferry". Together with his sister Louise, who was five years older, Ferry Porsche enjoyed a sheltered upbringing in his parents' house, where life naturally

der österreichischen Austro-Daimler-Werke arbeitete der Vater Ferdinand Porsche ununterbrochen an neuen Ideen und Konstruktionen. »Fest davon überzeugt, in einem Auto auf die Welt gekommen zu sein, konnte ich meinem Vater stundenlang zuhören, wie er über Automobile und Rennen sprach und aufregende Geschichten darüber erzählte,« erinnerte sich Ferry Porsche später. Die Welt der Maschinen beigeisterte ihn und so wurde das benachbarte Austro-Daimler-Werk zum bevorzugten Aufenthaltsort des Jungen: »Die Fabrik wurde mein Spielplatz. Mich erfüllten die vielen Wagen mit ihrer glänzenden Lackierung, die mit schimmernden Messinglampen und knolligen Hupen ausgerüstet überall herumstanden, mit fassungslosem Staunen.«

Ferry begann die väterlichen Automobile im Modellmaßstab nachzubauen und ließ sich dazu sogar im Werk Gussformen für Räder herstellen, die er dann eigenhändig mit Blei ausgoss. Was spielerisch so einfach schien, erwies sich in der Praxis jedoch als schwierig: Eine seiner ersten Fahrstunden auf dem Schoß des Vaters endete damit, dass er in die falsche Richtung lenkte und den jungfräulichen Kotflügel des

Im Alter von drei Jahren mit seinem Vater Ferdinand Porsche (1912)
With father Ferdinand Porsche, at the age of three (1912)

centered around the automobile. Father Ferdinand, as Chief Designer of the Austro-Daimler motor-vehicle company, worked without interruption on new ideas and designs. "I was quite convinced that I was brought into this world by car! I could listen to my father talking about cars and racing and telling the most exciting stories," Ferry Porsche recalled later. The boy was fascinated by the world of machines, loved to spend time in the neighboring Austro-Daimler factory. "It was my playground! I was speechless with amazement when I saw all those cars lined up with their highly polished paintwork, their gleaming brass lamps and their bulb horns."

Ferry soon began to make models of the cars his father had designed, and even persuaded the factory to make him some molds for casting the wheels, a task that he performed himself with molten lead. Driving looked equally simple, but his first attempt came to grief: sitting on his father's knees, he steered the car in the wrong direction, bent the fender of the Austro Daimler and damaged its pristine paint finish. Father Ferdinand was very angry, and young Ferry was obliged to switch to a less advanced form of

Ferry Porsche mit
seiner Schwester Louise
im Jahr 1915
*Ferry Porsche with sister
Louise in 1915*

transportation: he learned to ride a bike. Nonetheless, as these early years passed, his desire for a car of his own grew constantly. Most of his contemporaries would have had no chance of this wish becoming reality, but for the son of an already well-known automobile designer the situation was different: for Christmas 1920, the factory's apprentices built a small two-seater car – naturally designed by his father – with an air-cooled 6-horsepower twin-cylinder engine. It could reach the impressive top speed of 60 kilometers an hour. The surprise on Christmas Eve was complete: the parents had told their son that he would be getting a "goat pulling a cart". The new vehicle was henceforward known as the 'billy-goat car', and with it Ferry Porsche undertook many a lengthy tour of the local public streets. His conveyance had no license plates and he was much too young to tale a driving test, but as he admitted later "because of my father's position, the police in Wiener Neustadt always turned a blind eye."
With the same ease that he learned to ride a bicycle, Ferry Porsche very soon developed a talent for handling a motor vehicle quickly and safely. Only a short time after his eleventh birthday, he entered

Austro Daimler Zugwagen M 12 aus dem Jahr 1915. Zweiter Wagen von rechts: Auf dem linken Kotflügel Ferry Porsche, auf dem rechten Kotflügel seine Schwester Louise
Austro Daimler M 12 tractor, 1915. On the left fender of the second car from the right, Ferry Porsche; on the right fender, his sister Louise

Austro Daimler AD 617 Double Phaeton, aufgenommen vor der »Louisenhütte«, dem Jagdhaus von Ferdinand Porsche in Hochwolkersdorf. Am Lenkrad Ferry Porsche im Alter von 11 Jahren, rechts neben ihm seine Mutter Aloisia. Links am Fahrzeug Louise Porsche. Rechts neben Aloisia Porsche Ghislaine Kaes (1920)

Austro Daimler AD 617 Double Phaeton; the picture was taken in 1920 in front of Ferdinand Porsche's 'Louisenhütte' hunting lodge in Hochwolkersdorf. At the wheel, the 11-year-old Ferry Porsche; to the right, his mother Aloisia. Left, next to the car: Louise Porsche. At the right, next to Aloisia Porsche: Ghislaine Kaes (1920)

Austro Daimler verbog. Vater Ferdinand Porsche war darüber sehr ungehalten und bis auf weiteres musste sich Ferry auf ein anderes Fortbewegungsmittel konzentrieren: Er erlernte das Radfahren. Doch mit jedem Jahr wuchs der Wunsch nach einem eigenen Auto. Was für die meisten Gleichaltrigen nur ein Traum war, wurde für den Sohn des schon damals bekannten Automobilkonstrukteurs an Weihnachten 1920 zur Realität. In der Lehrlingsabteilung ließ der Vater einen kleinen Zweisitzer montieren, der, von einem luftgekühlten 6-PS-Zweizylindermotor angetrieben, stolze 60 Stundenkilometer erreichte. Groß war die Überraschung am Weihnachtstag, denn die Eltern hatten dem Sohn vorher erzählt, er bekomme einen Ziegenbock mit Wagen. Mit diesem »Ziegenbockwagen«, so der familieninterne Name des Gefährts, unternahm Ferry Porsche auf eigene Faust ausgiebige Fahrten auf öffentlichen Straßen. Zwar hatte das Wägelchen kein Kennzeichen und der Fahrer erst recht keinen Führerschein, doch »aufgrund der Stellung meines Vaters pflegten die Polizisten in Wiener Neustadt beide Augen zuzudrücken«, wie Ferry Porsche später bekannte. Wie zuvor das Radfahren hatte Ferry Porsche auch das Steuern eines Autos schnell und vor allem gut

LINKE SEITE: *Ferry Porsche 1920 am Steuer seines für ihn bei Austro Daimler gebauten Kinderautos*
LEFT PAGE: *Ferry Porsche in 1920, at the wheel of the kids' car built for him at Austro Daimler*

Mit dem »Ziegenbockwagen« während eines Automobil-Wettbewerbs in Wien (1921)
With the 'billy-goat car' during an automobile competition in Vienna (1921)

for a driving-skill tournament organized by the Vienna Automobile Club, and recorded the day's fastest time in his glorified kid's car. Ferdinand Porsche's experienced gaze soon registered his son's exceptional driving talent. In 1922, when the celebrated designer developed the 'Sascha' racing car on behalf of the industrialist Count Alexander "Sascha" Kolowrat, he surprised Ferry, then all of twelve years old, with the suggestion "You can take

»Mein Vater hat mich schon als Kind überall mitgenommen und dabei meinen überdurchschnittlichen Wissensdrang nie gebremst.«
Ferry Porsche und Vater Ferdinand (hinter dem linken Vorderrad) mit dem Austro Daimler »Sascha« 1922 beim Riesrennen in Graz/Steiermark

"As a child my father took me everywhere and never tried to suppress my urge to learn everything!"
Ferry Porsche and his father Ferdinand (behind the left front wheel) with the Austro Daimler 'Sascha' in 1922, at the Ries race meeting in Graz (Styria)

erlernt. Kurz nach seinem elften Geburtstag nahm er an einem Geschicklichkeitsturnier des Wiener Automobilclubs teil – und fuhr mit seinem Kinder-Wagen die schnellste Zeit des Tages. Dem geschulten Auge von Ferdinand Porsche blieb das fahrerische Talent seines Sohnes nicht verborgen. Als er 1922 den Rennwagen »Sascha« im Auftrag des Fabrikanten Alexander »Sascha« Graf Kolowrat entwarf, überraschte er den gerade mal zwölf Jahre alten Junior mit der Aufforderung: »So, jetzt kannst du den Wagen auf unserem Einfahrkurs einfahren.« Ferry Porsche war von dem 45 PS starken und 144 km/h schnellen Kleinwagen so begeistert, dass er sich fest vornahm, sich seinen eigenen Sascha zu bauen. Doch dazu kam es nicht mehr: Im März 1923 verließ Ferdinand Porsche Austro Daimler und wechselte als Technischer Direktor zur Daimler-Motoren-Gesellschaft nach Stuttgart.

Ferry Porsche im Alter von 13 Jahren mit seinem Vater Ferdinand Porsche (links neben ihm stehend, mit Schirmmütze) an der Rennstrecke in Monza (1922)
Ferry Porsche at the age of 13, with his father Ferdinand Porsche (standing on the left, wearing a peaked cap) at the Monza racetrack in 1922

the car out on our break-in track if you like!" Porsche junior was so delighted with this 45-hp small car, which was capable of a top speed of 144 kph, that he resolved to build a 'Sascha' of his own. This wish, however, was not to be fulfilled: in March 1923 Ferdinand Porsche left Austro-Daimler for Stuttgart, where he was appointed Technical Director of the Daimler-Motoren-Gesellschaft.

JUGENDZEIT IN STUTTGART
YOUTH IN STUTTGART

JUGENDZEIT IN STUTTGART
YOUTH IN STUTTGART

Auch für Ferdinand Porsche junior begann in Stuttgart ein neuer Lebensabschnitt, er musste sich auf das Großstadtleben einstellen. Ausflüge mit seinem »Ziegenbockwagen« waren nicht mehr möglich, denn in der schwäbischen Hauptstadt wurden die Straßenverkehrsgesetze streng geachtet. Vor allem aber musste er seine Spielkameraden zurücklassen, die aus allen Gesellschaftsschichten stammten. Zwar genoss er durch den prominenten Vater manche Vorteile, doch Standesdünkel war ihm stets fremd und so wurde der Sohn des Gärtners des elterlichen Anwesens sein bester Freund. In Wiener Neustadt war er vier Jahre zur Volksschule und anschließend zwei Jahre zur Realschule gegangen. Auf der Gottlieb-Daimler-Realschule in Cannstatt musste er sich neue Freunde suchen – manchen sollte er ein Leben lang verbunden

LINKE SEITE: Auf dem Hof der Porsche-Villa im Stuttgarter Norden (1924)
LEFT PAGE: *In the courtyard of the Porsche villa in the north of Stuttgart (1924)*

For Ferdinand Porsche junior too, a new phase in his life began in Stuttgart: he had to adjust to living in a large city. No more clandestine outings with his 'billy-goat car': in the capital of Germany's Swabian region, the rules of the road were taken seriously. He had to say goodbye to all his friends, who had so far been drawn from every social level. His father's important position brought Ferry many potential advantages, but he was never 'class-conscious', and his best friend was in fact the son of his parents' gardener. In Wiener Neustadt he had attended primary school for four years, followed by two years of secondary school. Now he was a pupil at the Gottlieb-Daimler secondary school in the Cannstatt suburb of Stuttgart and was obliged to make new friends – though some of these contacts proved to last a lifetime. Among his closest

bleiben. Zu seinen engsten Gefährten aus dieser Zeit zählte Albert Prinzing, der spätere Professor und Porsche-Geschäftsführer, der anlässlich von Ferry Porsches 75. Geburtstag die erste Begegnung wie folgt beschrieb: »Da kam ein junger Österreicher mit langem Haar in eine Schulklasse, die ›Hindenburgschnitt‹ trug; er hatte uns unbekannte Hosen an, die er Knickerbockers nannte. Schon damals war er das, was man heute einen Trendsetter nennt, denn bald wurden unsere Haare länger und die Knickerbockers Mode in der Klasse.« Zeit seines Lebens war Ferry Porsche für seinen feinen und ausgesuchten Geschmack genauso bekannt wie für seine Stetigkeit und Beständigkeit bei menschlichen Beziehungen. Genauso prägend wie die neue Schule sollte für Ferry Porsche auch das im Dezember 1923 bezogene neue Domizil der Familie werden. Die von dem bekannten Architekten Paul Bonatz im Stuttgarter Norden erbaute Villa Porsche wurde zum Stammsitz der Familie und – neben dem Schüttgut im österreichischen Zell am See – zu einem Zentrum im Leben von Ferry Porsche, in dem Automobilgeschichte geschrieben werden sollte.

Jugendportrait von Ferry Porsche (um 1923)
A youthful portrait of Ferry Porsche (about 1923)

acquaintances from that period was Albert Prinzing, who was later to be granted a Professor's title and become the Porsche company's General Manager. On the occasion of Ferry Porsche's 75th birthday, Prinzing recalled their first meeting in the following terms: "One day an Austrian boy with long hair joined our class. We all had our hair cut in 'Hindenburg style'. The newcomer wore pants that were totally unknown to us, which he called 'knickerbockers'. He was evidently something of a trend-setter: in no time at all we began to grow our hair longer and 'knickerbockers' were all the fashion in our class." Ferry Porsche was noted throughout his life for his excellent, reliable taste, as indeed for his consistency and trustworthiness when it came to human relationships. Not only the new school had its influence on Ferry Porsche's adolescent years, but also the family's new home: the Porsche villa in the north of Stuttgart, designed by the well-known architect Paul Bonatz, into which the Porsches moved in December 1923. In addition to the 'Schüttgut' estate in Zell am See, Austria, the new house began to play a central part in Ferry Porsche's life, and was also destined to become a place in which automobile history was created.

Wie schon zuvor in Wiener Neustadt ließ Ferdinand Porsche seinen Sohn auch in Stuttgart an seiner Arbeit teilhaben. Dank einer Sondergenehmigung erhielt Ferry Porsche mit nur 14 Jahren den Motorrad- und mit 16 Jahren einen Kraftfahrzeugführerschein und durfte fortan alle Prototypen fahren, die sein Vater aus dem Daimler-Werk in Untertürkheim mit nach Hause brachte. Das amtliche Dokument hatte aber auch weitere Vorteile für den Heranwachsenden: »Wenn ich damals ins Kino ging, um einen nicht jugendfreien Film anzusehen, so zeigte ich nur schnell meinen Führerschein, und ich konnte hineingehen.« Am liebsten sah er Filme mit Charlie Chaplin, Buster Keaton oder Harold Lloyd, die er schon bald zu parodieren verstand und damit zeitlebens so manche Gesellschaft zum Lachen brachte. Den Mittelpunkt seines Lebens bildete jedoch die Begeisterung für Automobile. Auch auf längeren Fahrerprobungen in den Schwarzwald begleitete er den Vater und beeindruckte Passanten am Steuer der mächtigen Mercedes-Kompressormodelle. Mit 18 Jahren schließlich legte Ferry Porsche die reguläre Führerscheinprüfung ab und bekam ein eigenes Motorrad, eine 500er BMW. Allerdings verblasste der Reiz des Zweirades schnell neben einer anderen Erfahrung,

Auf seinem BMW-Motorrad auf dem Hof der Porsche-Villa (1927)
On his BMW motorcycle in the courtyard of the Porsche villa (1927)

Just as he had in Wiener Neustadt, Ferdinand Porsche involved his son actively in his work for the new employer in Stuttgart. He succeeded in obtaining a special license for Ferry to ride a motorcycle at the age of 14 and, two years later, to drive a car. From then on, the son drove all the prototypes that his father brought home from the Daimler factory in Untertürkheim. The official document had other advantages too for the young man: "If I wanted to see an adult movie, I gave them a quick glance at my driver's license, and they always let me in." But Ferry's favorite movies starred Charlie Chaplin, Buster Keaton and Harold Lloyd; before long he was imitating them perfectly, and years later was always able to raise gales of laughter from party guests. But even then, his central interest in life was enthusiasm for the automobile. He often accompanied his father on extended journeys to the Black Forest, and never failed to impress passers-by when he took the wheel of a mighty supercharged Mercedes. As soon as he reached the age of 18, Ferry Porsche passed the regular driving test. He was also given a motorcycle of his own, a 500-cc BMW, but the pleasure of riding this bike soon paled: in September 1927 he met and

Ferry Porsche, Schwester Louise und seine spätere Ehefrau Dorothea (geb. Reitz) bei einer Ausfahrt mit einem Mercedes-Benz Typ 24/100/140 im Winter 1927/28
Ferry Porsche, his sister Louise and his later wife Dorothea (née Reitz) on a winter outing in a Mercedes-Benz Type 24/100/140 (1927/28)

Ferry Porsche im Alter von 18 Jahren (gebückt mit Mütze) mit seinem Vater Ferdinand Porsche mit einem Mercedes-Benz Typ 5/25 am Wörthersee
Ferry Porsche at the age of 18 (bending down, wearing a cap) with his father Ferdinand Porsche at the Wörthersee Lake; the car is a Mercedes-Benz Type 5/25

die ein Leben lang anhalten sollte: Im September 1927 verliebte er sich in die Stuttgarterin Dorothea Reitz, die er 1935 heiratete und mit der er bis zu ihrem Tod im Jahr 1985 glücklich zusammen lebte. Die beiden wurden Eltern der vier Söhne Ferdinand Alexander (*1935), Gerhard (*1938), Hans-Peter (*1940) und Wolfgang (*1943), die als Gesellschafter der heutigen Dr. Ing. h.c. F. Porsche AG das Lebenswerk von Ferry Porsche weiterführen.

Nach dem Abschluss der Mittleren Reife stand für Ferry Porsche fest, dass er als Automobilkonstrukteur in die Fußstapfen seines Vaters treten wollte: »Je mehr ich das Leben verstehen lernte, desto mehr bewunderte ich das leuchtende Beispiel, das mir mein Vater gab.« Ferry Porsche erhielt 1928 die damals noch seltene Chance, für ein Jahr bei der damaligen Robert Bosch AG als Praktikant zu arbeiten. Dort lernte er Manfred Behr kennen, dessen Familie Autokühler herstellte und der zu einem seiner engsten Freunde werden sollte. Dieser berichtete anlässlich des 60. Geburtstages von Ferry Porsche: »Bosch hat damals zum ersten Mal Praktikanten aufgenommen. Zwölf Bewerber wurden auf Grund von Eignungsprüfungen heraus-

Ferry Porsche (1927)
Ferry Porsche, 1927

fell in love with Dorothea Reitz, who also lived in Stuttgart. They married in 1935, and lived contentedly together until her death in 1985. They had four sons: Ferdinand Alexander (born in 1935), Gerhard (born in 1938), Hans-Peter (born in 1940) and Wolfgang (born in 1943). As partners in the current Dr. Ing. h.c. F. Porsche AG company, they continue Ferry Porsche's life's work.

After obtaining the first secondary school certificate, Ferry Porsche was determined to follow in his father's footsteps and become an automobile designer. "The morc I began to understand what life was all about, the more I admired the splendid example set by my father." In 1928, Ferry Porsche was given an opportunity not open to many young people at that time. He joined what was then Robert Bosch AG as a student apprentice. At that company he made the acquaintance of Manfred Behr, whose family manufactured automobile radiators. Behr, who became one of his closest friends, was quoted as follows when Ferry Porsche celebrated his 60th birthday: "That was the first time that Bosch had taken on any student apprentices. Twelve candidates were examined

gesucht. Ein volles Jahr musste man sich verpflichten. 6.35 Uhr morgens war damals Arbeitsanfang. Wenige Minuten später wurde das Werkstor geschlossen. Zuspätkommende, egal ob Arbeiter, Angestellter, Lehrling oder Praktikant, wurden nicht mehr eingelassen. Sie waren praktisch ausgesperrt. Damals – bei fünf bis sechs Millionen Arbeitslosen – wurde nur an zwei bis drei Tagen gearbeitet. Und mehr als fünf Minuten Zuspätkommen bedeutete den Lohnausfall von einem ganzen Tag.« Nach dem Abschluss des einjährigen Praktikums folgte er seinem Vater nach Österreich, nachdem dieser die Daimler-Benz AG verlassen und Anfang 1929 eine Position als Chefkonstrukteur der Steyr-Werke AG angenommen hatte.

Zur Vorbereitung eines technischen Studiums besuchte Ferry Porsche in Wien eine Privatschule, doch verbrachte er bald mehr Zeit in den Konstruktionsbüros und Werkstätten des Vaters, als auf der Schulbank. Als Ferdinand Porsche 1930 die Steyr-Werke wieder verließ, stand für den Sohn fest, dass er anstatt eines Studiums eine praktische Ausbildung an der Seite des Vaters absolvieren wollte. Aber inmitten der großen Weltwirtschaftskrise war die wirtschaftliche Zukunft der Familie Porsche ungewiss.

Im Alter von 20 Jahren vor dem Wohnhaus der Familie Porsche in Steyr-St. Ulrich
At the age of 20, in front of the Porsche family's house in Steyr-St. Ulrich

and found to be suitable. They had to sign up for a complete year. Work started at 6.35 every morning, and a few minutes later, the factory gate was closed. Nobody who arrived later than that – whether they were a humble worker, a member of the staff, a trainee or a student apprentice – was allowed in. At that time, Germany had between five and six million unemployed, and the factory only worked two or three days a week. Anyone who arrived more than five minutes too late wasn't paid for that day." After completing the year's student apprenticeship, Ferry Porsche followed his father back to Austria. The latter had left Daimler-Benz AG and, early in 1929, been offered the position of Chief Designer at Steyr-Werke AG.

To prepare for his engineering studies, Ferry Porsche attended a private school in Vienna, but before long he was spending more time in his father's design offices and workshops than in the lecture rooms. In 1930, when Ferdinand Porsche left Steyr again, his son was determined not to study any further but to learn his trade in a practical way alongside his father. Then came the monumental world economic crisis, however, and suddenly the Porsche family's future no longer looked secure. "My father had to ask

LINKE SEITE:
Ferry Porsche mit einem Steyr-Prototyp (1929)
LEFT PAGE:
Ferry Porsche with a Steyr prototype, 1929

LINKS: 1929: Ferry Porsche mit einem von seinem Vater Ferdinand Porsche entwickelten Steyr Typ 30
ON THE LEFT: *1929: Ferry Porsche with a Steyr Type 30 developed by his father Ferdinand Porsche*

RECHTS: Zeitungsartikel vom Dezember 1930 über die Eröffnung eines unabhängigen Konstruktionsbüros durch Ferdinand Porsche
ON THE RIGHT: *A newspaper article dating from 1930 on the opening of Ferdinand Porsche's independent design office*

»Für meinen Vater erhob sich die Frage: Was sollte er in Zukunft tun? Er hatte zahlreiche neue Angebote. Aber in seinem Kopf reifte eine andere Idee. Er hatte langsam genug davon, für andere zu arbeiten.« Über die genauen Motive, künftig selbstständig arbeiten zu wollen, berichtete Ferry Porsche im Jahr 1981: »Mein Vater hatte erkannt, dass seine Produktivität so enorm war, dass er innerhalb von zwei bis drei Jahren einer Firma so viel Grundlagen geschaffen hat, dass sie zehn Jahre davon leben konnte. Und die Firma hat sich dann die Ausgaben gespart, in dem sie ihm den Vertrag nicht verlängert haben. Und er hat sich gesagt: Wie komme ich dazu jedes Mal Grundlagen zu schaffen und dann kann ich wieder gehen. Es ist doch besser, ich mache ein eigenes Büro und arbeite für alle.« Das Resultat dieser Überlegungen war Ende 1930 in einer Stuttgarter Zeitung zu lesen: Dr. Ferdinand Porsche beabsichtige, nach Stuttgart zurückzukehren, um dort ein eigenes Konstruktionsbüro zu eröffnen.

An der Rennstrecke (1930)
At the race track, 1930

RECHTE SEITE: Im Garten der Porsche-Villa (1931)
RIGHT PAGE: *In the garden of the Porsche villa, 1931*

himself what his next move should be. He had several offers from other companies, but this was when he conceived a totally different idea. He was frustrated by having to work for other people." In 1981 Ferry Porsche summed up the precise motives that led to his father going it alone, as a self-employed person, in the following words: "My father realized that his productivity was so great that within two to three years he could provide his employers with the resources to survive for another ten, after which the company would save money by not renewing his contract. And so he said to himself: 'Is it my fate in life to create value for others, only to have them get rid of me? I would do better to open an office of my own and work for everyone!'" By the end of 1930 his plans were so far advanced that the local newspaper, the Stuttgarter Zeitung, was able to write: 'Dr. Ferdinand Porsche intends to return to Stuttgart and open his own design office here.'

LEHRJAHRE IM KONSTRUKTIONSBÜRO
EARLY YEARS IN THE DESIGN OFFICE

LEHRJAHRE IM KONSTRUKTIONSBÜRO
EARLY YEARS IN THE DESIGN OFFICE

Am 25. April 1931 wurde die »Dr. Ing. h.c. F. Porsche GmbH – Konstruktion und Beratung für Motoren- und Fahrzeugbau« ins Stuttgarter Handelsregister eingetragen. Gesellschafter waren neben Ferdinand Porsche sein Schwiegersohn Anton Piëch und der ehemalige Rennfahrer Adolf Rosenberger. Das Arbeitsspektrum des zu Beginn zwölfköpfigen Teams um Ferdinand Porsche umfasste die gesamte Bandbreite der Kraftfahrzeugtechnik. Ferry Porsche war dabei von Beginn an in alle Projekte und Auftragsentwicklungen des Konstruktionsbüros einbezogen. Als jüngster Mitarbeiter absolvierte Ferry Porsche zunächst eine Lehrzeit in den Bereichen Konstruktion und Versuch. Sein Lehrer war der Diplom-Ingenieur Walter Boxan, der ihm die theoretischen und rechnerischen Grundlagen der Automobiltechnik vermittelte. 1931 entwickelten die

LINKE SEITE: Ferry Porsche im Stuttgarter Konstruktionsbüro, Kronenstraße 24
LEFT PAGE: Ferry Porsche in the Stuttgart design office on Kronenstrasse 24

Visitenkarte der
Dr. Ing. h.c. F. Porsche GmbH
*Visiting card of
Dr. Ing. h.c. F. Porsche GmbH*

On April 25, 1931 details of a new company appeared in the Stuttgart trade register. Its name: 'Dr. Ing. h.c. F. Porsche GmbH – Konstruktion and Beratung für Motoren- and Fahrzeugbau' (Engine and Vehicle Design and Consultation). The partners, in addition to Ferdinand Porsche himself, were his son-in-law Anton Piëch and the former racing driver Adolf Rosenberger. Ferdinand Porsche had assembled a team of twelve employees, ready to apply their talents to every aspect of motor-vehicle technology. Ferry Porsche was involved from the very start in all the new design office's projects and the development work it was commissioned to perform. As the new company's youngest employee, Ferry Porsche was first of all allocated to the Design and Experimental departments as a trainee. He was taught there by

Porsche-Ingenieure für den Chemnitzer Automobilhersteller Wanderer eine neue Mittelklasselimousine. Bei diesem Projekt erledigte der 21-jährige schon eigene Aufgaben, wie die Verbesserung der Lenkung für den neuen Zweiliter-Wagen. Ein Lehrstück, das sich als unerwartet ergiebig erwies, denn die Lenkung wurde in weiterentwickelter Form später auch für den Auto-Union-Rennwagen und für den Volkswagen genutzt.

Im Konstruktionsbüro reihte sich Projekt an Projekt: eine Schwingachse für die Horch-Werke Zwickau und ein im Auftrag der Zittauer Phänomen-Werke konstruierter Fünfzylinder-Sternmotor entstanden ebenfalls noch im selben Jahr. Zu einem Meilenstein wurde die am 10. August 1931 zum Patent angemeldete Drehstabfederung, die über viele Jahrzehnte im internationalen Automobilbau zum Einsatz kam. Im Auftrag von Zündapp entwarf das Porsche-Konstruktionsbüro 1932 einen Kleinwagen mit der internen Bezeichnung Typ 12, der dem Nürnberger Zweiradhersteller aus der Motorrad-Krise Anfang der 30er Jahre helfen sollte. Das Projekt wurde mit dem Wiederaufleben des Motorradmarktes eingestellt, aber die Arbeit war alles andere als vergebens: Erstmals manifestierte sich

LINKE SEITE: 1932 im Garten der Villa Porsche: (v.l.n.r.) Ferry Porsche, Aloisia Porsche, Waltraud Reitz, Ferdinand Porsche, Karl Rabe, Heidi Rabe, Anna Glaser, Dorothea Porsche, Herbert Kaes
LEFT PAGE: *In the garden of the Porsche villa in 1932: (left to right) Ferry Porsche, Aloisia Porsche, Waltraud Reitz, Ferdinand Porsche, Karl Rabe, Heidi Rabe, Anna Glaser, Dorothea Porsche and Herbert Kaes*

Fahrgestell und Motor des Wanderer W21 (Porsche Typ 7) aus dem Jahr 1931
Chassis and engine of the 1931 Wanderer W21 (Porsche Type 7)

Fahrgestell und Motor des Wanderer W21 (Porsche Typ 7) aus dem Jahr 1931
Chassis and engine of Wanderer W21 (Porsche Type 7) from 1931

RECHTE SEITE:
Ferry Porsche und das für Zündapp entwickelte Typ 12 Cabriolet (um 1932)
1934 erhielt Ferry Porsche (rechts im Bild) die Verantwortung für die Durchführung der Versuchsfahrten

RIGHT PAGE:
Ferry Porsche and the Type 12 convertible developed for Zündapp (about 1932)
In 1934 Ferry Porsche (at right in picture) assumed responsibility for test driving

das technische Konzept des späteren Volkswagens. Ferry Porsche sammelte mit dem Prototyp wertvolle Erfahrungen – persönlicher und technischer Natur. So blockierte bei einer Probefahrt auf der Solitude-Rennstrecke bei Stuttgart, die alltags Teil des öffentlichen Straßennetzes war, in einer Rechtskurve die Lenkung – eben jene, die er ursprünglich für den Wanderer selbst konstruiert hatte. »Um eine möglichst billige Lenkung zu bekommen, hatten wir versucht, die Lenkspindel nicht wie üblich in Rollen- oder Kugellagern zu führen,

Walter Boxan, a graduate engineer who instructed him in the theory and basic calculations needed in automobile technology. In 1931, Porsche's engineers developed a new midsize sedan for Wanderer, an automobile manufacturer based in Chemnitz. The 21-year-old Ferry Porsche was allowed to assume responsibility for some of the work on this new two-liter car, for example improvements to the steering. This task proved to be of lasting value: in further developed form, this steering system was used later on the Auto Union racing cars and on the Volkswagen, the 'People's Car'.

The design office was overwhelmed with work: order of all kinds began to pour in. A swing-axle suspension layout was designed for Horch in Zwickau, and in the same year a five-cylinder radial engine was commissioned by the Phänomen-Werke in Zittau. A notable milestone was Porsche's torsion-bar vehicle suspension system, for which a patent was applied on August 10, 1931, and which remained in use throughout the international motor-vehicle industry for many decades afterwards. In 1932, the Porsche design office drafted out a small-car design, known

LEHRJAHRE IM KONSTRUKTIONSBÜRO | EARLY YEARS IN THE DESIGN OFFICE

sondern in Gleitlagern. Bei dieser Einbauweise lag aber das obere Lager über dem Ölniveau, lief trocken und saß fest. Um nun keine fremde Hilfe in Anspruch nehmen zu müssen, half ich mir folgendermaßen: Ich löste die Befestigung des oberen Lagers. So war es möglich, nach einer Seite zu lenken.« Die Heimfahrt dauerte entsprechend lang, denn bei jeder Kurve in eine andere Richtung musste der Versuchsfahrer aussteigen und an den Rädern von Hand den notwendigen Winkel einschlagen. Und er erfuhr am eigenen Leib was es bedeutet, wenn ein Konstrukteur am falschen Ende spart. Mit der Zeit wurden die Aufgaben von Ferry Porsche immer verantwortungsvoller, so dass er ab 1932 die Koordination der Konstrukteure leitete, die Versuche überwachte und zusammen mit dem Vater die Verbindungen zu den Auftraggebern pflegte.

Im Frühjahr 1933 erhielt das Konstruktionsbüro Porsche von der sächsischen Auto Union den Auftrag, nach den Regeln der neuen 750-kg-Rennformel einen 16-Zylinder-Rennwagen zu entwickeln. Unmittelbar nach Vertragsabschluss begann die Porsche-Mannschaft unter der Leitung von Oberingenieur Karl Rabe die Arbeit an dem als Mittelmotor-Fahrzeug ausgelegten Auto Uni-

Portrait (um 1935)
Portrait (about 1935)

internally as the Type 12, for the Zündapp company in Nuremberg; it was intended to help this manufacturer of motorcycles to survive the crisis on its market in the 1930s. The project was abandoned when the motorcycle market recovered, but the work was by no means in vain: this was the first manifestation of the basic technical concept that was later to be applied to the Volkswagen. With this prototype, Ferry Porsche was able to amass a vast amount of experience – both personal and technical. During a test run on the Solitude racetrack near Stuttgart, which consisted of public roads that were only closed when a race meeting took place, the car's steering jammed while the car was taking a right-hand corner. This was basically the same steering system that Ferry Porsche had designed himself for the Wanderer sedan. "To keep the cost of the steering down, we had used plain bearings for the steering column instead of the usual ball or roller bearings. We failed to see that the upper bearing could be above the oil level in the steering box, which would cause it to run dry and seize." To avoid having to call for outside help, Ferry Porsche tackled the problem as follows: "I unscrewed the upper bearing mount. I was then able to steer

Chassis des Auto Union
Grand-Prix-Rennwagen
(Porsche Typ 22)
Chassis of the Auto Union
Grand Prix racing car
(Porsche Type 22)

Bernd Rosemeyer im Typ 22 beim Training zum Vanderbilt Cup 1937. Links Ferry Porsche
Bernd Rosemeyer in the Type 22 during practise for the 1937 Vanderbilt Cup. At left: Ferry Porsche

Vater und Sohn (1936)
Father and son, 1936

on P-Rennwagen (P für Porsche). Ende 1933 lief bereits der erste Motor auf dem Prüfstand – und bereitete erst einmal Probleme. Eines davon war die Kurbelwelle, deren vorderes Ende immer blau anlief. Ferry Porsche tippte auf die unterschiedliche Wärmeausdehnung von Stahlkurbelwelle und Elektron-Motorgehäuse und infolgedessen auf ein zu geringes Längsspiel. »Die Ingenieure hörten mich höflich an, waren aber nicht überzeugt, dass ich recht haben konnte«, ärgerte sich

the car at least in one direction!" The journey home took some time to complete, because whenever the car approached a corner in the 'wrong' direction, the unfortunate test driver had to climb out and turn the wheels by hand to approximately the correct angle. In this way he was directly confronted with the consequences of attempting to save money in the wrong place. As time went by, Ferry Porsche was entrusted with increasingly important tasks, and from 1932 on was responsible for design coordination, supervising tests and, together with his father, maintaining close contact with clients.

In the spring of 1933 the Porsche design office was contacted by the Auto Union carmaking group, which was based in the German state of Saxony, and asked to develop a racing car to comply with the then current 750-kg weight formula. As soon as the ink was dry on the contract, a Porsche team led by Senior Engineer Karl Rabe started work on what was to be a 16-cylinder, mid-engine design. By the end of 1933, the first engine for the Auto Union P (for Porsche) racing car was undergoing rig tests – and giving rise to a number of problems.

Ferry Porsche. »Ohne weitere Diskussion führte ich einen eigenen Test durch, nahm die Kurbelwelle und das Kurbelgehäuse und ging damit in die Härterei, wo entsprechende Öfen zur Erwärmung der beiden Teile zur Verfügung standen.« Die anschließenden Messungen bestätigten seine Theorie, und von nun an wurde mehr »Endspiel« eingeplant. Anfang 1934 fanden die ersten Versuchsfahrten mit dem Grand-Prix-Wagen statt, bei denen Ferry Porsche sein fahrerisches Talent so eindrucksvoll unter Beweis stellte, dass der Vater Ferdinand künftige rennsportliche Ambitionen seines Sohnes befürchtete. Mit dem Hinweis auf eine vielversprechende Zukunft als Automobilkonstrukteur verbot er dem Sohn weitere Fahrten im Rennwagen mit den Worten: »Rennfahrer habe ich viele, aber nur einen Sohn!« Ferry Porsche fügte sich und lebte seine sportlichen Ambitionen bei Rallyes mit einem Wanderer-Tourenwagen aus. Bei der 2.000 Kilometer-Fahrt durch Deutschland fuhr er in direkter Konkurrenz zu den damaligen Top-Fahrern Bernd Rosemeyer, Hans Stuck und Prinz zu Leiningen. »Im ersten Abschnitt der Rallye, der durch den Schwarzwald führte, war meine Zeit sogar besser als die von Rosemeyer«, triumphierte Ferry Porsche noch viele Jahre später.

LINKE SEITE: Ferry Porsche am Steuer eines offenen Wanderer-Sport
LEFT PAGE: *Ferry Porsche at the wheel of an open Wanderer Sport model*

Ferry Porsche (mitte), Herbert Kaes (links) und Motorenmeister Erwin Kenner an einem 4-Zylinder-Viertakt-VW-Motor in der Garage der Villa Porsche Stuttgart-Nord (1936)
Ferry Porsche (center), Herbert Kaes (left) and master engine mechanic Erwin Kenner with a four-cylinder, four-stroke VW engine in the garage of the Porsche villa in the north of Stuttgart (1936)

There was blue discoloration, for instance, at the front end of the crankshaft. Ferry Porsche attributed this to inadequate longitudinal clearance caused by differences in thermal expansion between the steel crankshaft and the engine block itself, which was cast in 'elektron', a magnesium alloy. Ferry Porsche was irritated: "The engineers listened politely, but were evidently not convinced that what I said was true! So without any further discussion I held a test of my own: I took the crankshaft and the engine block to the hardening shop, where there were furnaces capable of heating these two large components." After the test, measurements confirmed Ferry Porsche's view, and more endplay was therefore provided on later engines. The new Grand Prix car was rolled out for initial racetrack testing at the beginning of 1934, and Ferry Porsche demonstrated his skill at the wheel so impressively that father Ferdinand has every reason to fear that his son could seek new stardom as a racing car driver. He forbade this by pointing out what a splendid career awaited him as a car designer, and ending with the words "I can get any number of racing drivers, but I only have one son!" Ferry Porsche agreed, and from then on satisfied his

1934: DAS VOLKSWAGEN-PROJEKT

Der legendäre Auto Union 16-Zylinder-Rennwagen war nur der Auftakt zu einem weiteren Erfolg im Jahr 1934: dem Volkswagen. Am 17. Januar präsentierte Ferdinand Porsche ein »Exposé betreffend den Bau eines deutschen Volkswagens« in dem er sein Konzept für einen robusten und preisgünstigen Kompaktwagen vorstellte. Aber der Weg zum Entwicklungsauftrag war mühevoll, zumal die etablierten deutschen Hersteller den Volkswagen als potenziellen Konkurrenten ablehnten. Auf politischen Druck schloss der Reichsverband der Automobilindustrie (RDA) am 22. Juli 1934 mit Porsche einen Vertrag über den Bau eines Volkswagen-Prototypen ab. Zwei Bedingungen machten diese Aufgabe besonders problematisch: Der anvisierte Kaufpreis von weniger als 1.000 Reichsmark und der Termin für die Fertigstellung des Prototyp in nur zehn Monaten. Ferry Porsche stufte später die Entwicklung des Auto Union Rennwagens im Vergleich zum Volkswagen als »Kinderspiel« ein. Denn beim Volkswagen mussten die Ingenieure nicht nur konstruieren, sondern von Anfang an auch kalkulieren. Es galt einen Endpreis von 990 Mark einzuhalten, zu dem der Volkswagen

»Der Käfer ist unser Kind. Ein altes Kind, ohne Zweifel, aber aus vornehmem Hause.«
Ferry Porsche mit seiner Ehefrau Dorothea 1936 im zweiten Volkswagen-Prototypen »V2« auf dem Marktplatz in Tübingen

"The Beetle is our child. Agreed, it's not as young as it used to be, but it comes from a fine family."
Ferry Porsche with his wife Dorothea in the second Volkswagen prototype ('V2') on Market Square in Tübingen (1936)

motor-sport ambitions by rallying a Wanderer tourer. On the German 2000-kilometer run he found himself competing directly against Bernd Rosemeyer, Hans Stuck and Prinz zu Leiningen, the top drivers of their day. Many years later he was still able to recall with pride: "In the first section of the rally, through the Black Forest, I put up a better time than Rosemeyer!"

1934: THE PEOPLE'S CAR PROJECT

The legendary Auto Union 16-cylinder racing car was followed by a further success in 1934: the Volkswagen. On January 17 Ferdinand Porsche presented an 'Exposé on the Construction of a German People's Car', his concept for a strongly built, low-priced compact car. This was the start of a long and arduous struggle to obtain a development contract, with well-established German carmakers firmly opposed to a project that would result in a potential rival sure to harm their sales. None the less, political pressure on the Automobile Industry Federation of the German Reich (RDA) obliged it to conclude a contract with Porsche on July 22, 1934 for the construction of a People's Car prototype. Two conditions were

verkauft werden sollte. So musste zum Beispiel auf eine hydraulische Bremse verzichtet werden, da Lizenzgebühren für den Patentinhaber Lockheed angefallen wären. »Das Weglassen war das Entscheidende. Wir sind ganz systematisch vorgegangen – der Radstand ergab sich aus dem Raum, den vier Erwachsene mit akzeptablem Platzangebot benötigten. Die Spurweite wurde so gewählt, dass der Wagen auch auf Feldwegen und durch schmale Dorfdurchfahrten kommen konnte.«

Die Zeit drängte, und auch die knappen finanziellen Mittel wirkten nicht projektfördernd, denn der Entwicklungsvorschuss von 20.000 Reichsmark monatlich erwies sich schnell als viel zu niedrig. Die Konstrukteure waren gezwungen, mit Bau und Montage der ersten Versuchswagen in der Garage der Porsche-Villa in Stuttgart zu beginnen. Platz war Mangelware, zumal der Maschinenpark das Raumangebot in Ferry Porsches Privatwerkstatt noch zusätzlich strapazierte. Zu Bohr- und Fräsmaschine kamen noch zwei Drehbänke und die zwölfköpfige Entwicklungsmannschaft. »Fragen Sie mich nicht, wie wir es machten«, erinnerte er sich, »aber die ersten drei Prototypen, VW Serie 3 genannt, wurden dort gebaut.« Die Entwicklung des Volkswagen

Alpenversuchsfahrt vom 10. bis 17. September 1936 mit zwei »V3«-Prototypen: (V.l.n.r.) Ferdinand Porsche, Ferry Porsche, Kfz-Meister Rudolf Ringel, Chauffeur Josef Goldinger
Test driving in the Alps, from September 10 to 17, 1936 with two 'V3' prototypes: (from left to right) Ferdinand Porsche, Ferry Porsche, car mechanic Rudolf Ringel and driver Josef Goldinger

included that made the project very much less likely to succeed: the target selling price was not to exceed a thousand Reichsmarks, and the prototype was to be completed within ten months. Later, Ferry Porsche was to refer to development of the Auto Union racing car as 'child's play' compared with the task of completing the first Volkswagen prototype. For this project, the engineers not only had to design everything from scratch but also keep a careful eye from the start on the total cost, which had to be below the 990 RM target at which the Volkswagen was to be sold. Hydraulic brakes, for instance, were out, since they would have meant paying a license fee to Lockheed, the patent holder. "Most of our decisions concerned what to leave out! We approached the task systematically: the wheelbase was determined by the need to provide an acceptable amount of space for four adults. The track was not too wide, to permit the car to be driven on rural roads and through narrow village streets."

Time was short, and insufficient funds had been allocated for the project to be pushed ahead as it deserved to be: the monthly payments of

Ferdinand und Ferry Porsche während ihrer Reise in die USA (1937)
Ferdinand and Ferry Porsche during their journey to the USA (1937)

RECHTE SEITE: Ferry und Dorothea Porsche auf dem Schnelldampfer »Bremen« (1937)
RIGHT PAGE: *Ferry and Dorothea Porsche on board the steamship 'Bremen' (1937)*

dauerte unterdessen länger als geplant. Fast genau ein Jahr nach dem offiziellen Entwicklungsauftrag war der erste Volkswagen, der V1 (V = Versuchswagen), fahrfertig. Am 3. Juli 1935 stellte Ferdinand Porsche die Limousine einer Kommission des RDA vor. Der zweite Versuchswagen, ein Cabriolet mit dem Namen V2, trat am 22. Dezember zur Jungfernfahrt an. Zwei Monate später, am 24. Februar 1936, feierten die beiden ersten Volkswagen offiziell Weltpremiere in Berlin. Im Umfeld der hochkarätigen Porsche-Ingenieure hatte

20.000 Reichsmarks soon proved to be totally inadequate. The designers economized wherever they could: for instance, assembly of the first experimental cars took place in the garage of the Porsche villa in Stuttgart. Space was limited, and the machines soon began to squeeze Ferry Porsche's private workshop into a corner. As well as a power drill and a milling machine, the twelve-strong development team needed two lathes. Ferry Porsche's comment later: "Don't ask me how we did it, but we managed to build the first three prototypes, our VW Series 3, in that confined space." Development of the Volkswagen took longer than planned, but the first Volkswagen, the V1 (V was short for the German word for 'Experimental') was ready to drive just under a year after the official development contract had been signed. Ferdinand Porsche demonstrated the sedan to an RDA commission on July 3, 1935. The second experimental car, a convertible known as V2, took to the road on December 22, and two months later, on February 24, 1936, the first two Volkswagens were premiered officially in Berlin.

Ferry Porsche had learned a lot from the company's highly qualified engineers and was no longer a trainee

Ferry Porsche viel gelernt und sich vom Praktikanten zum anerkannten Junior-Chef entwickelt. Ferdinand Porsche förderte und forderte den Sohn, als er ihm 1935 die Leitung der Fahrerprobung des Volkswagens übertrug. Bis zum Herbst 1936 entstanden die ersten V3-Prototypen, mit denen eine systematische Fahrerprobung durchgeführt wurde. Als Versuchsleiter übernahm Ferry Porsche die Aufgabe, bis zum Jahresende 50.000 Kilometer Testfahrt zurückzulegen. Im Endspurt auch mit Sonntagsschichten gelang es dem Versuchsteam, alle drei Autos bis zum 22. Dezember 1936 über die gewünschte Distanz zu bringen. Für Ferry Porsche war dieser erste Test nicht nur eine technische, sondern auch eine politische Aufgabe. Zwar war er Versuchsleiter des Konstruktionsbüros Porsche, aber der Reichsverband der Automobilindustrie hatte Mitarbeiter zur kritischen Überwachung der Versuche in das Team geschickt. Bald gab es unterschiedliche Auffassungen über die Testergebnisse. Doch am Ende fiel der einhundert Seiten starke Bericht des RDA positiv aus: »Das Fahrzeug hat Eigenschaften gezeigt, die eine Weiterentwicklung empfehlenswert erscheinen lassen.« Entgegen der ersten Überlegung, den Volkswagen

Auf Versuchsfahrt mit dem VW-Prototyp »V303« Rolldachlimousine (1938)
A test run with the VW 'V303' prototype (1938)

RECHTE SEITE: Ferry Porsche im »V303«-Cabriolet (1938)
RIGHT PAGE: *Ferry Porsche in the 'V303' convertible (1938)*

but a universally accepted junior partner. Ferdinand Porsche encouraged his son but imposed tough demands on him as well, for instance in 1935 when he put him in charge of Volkswagen road testing. The first V3 prototypes were completed in the fall of 1936, and were immediately subjected to systematic road trials. Ferry Porsche's task was to complete a 50,000-kilometer driving program by the end of the year. As time grew short, the team worked seven days a week, and by December 22, 1936 all three cars had covered the required distance. For Ferry Porsche, this initial test was as much a political challenge as a technical one. Although he supervised the driving program on behalf of the Porsche design office, the Automobile Industry Federation sent its own critical observers to accompany the team. Before long, the test results were being interpreted in two different ways. Despite this, the RDA's concluding report, a hundred pages long, reached a positive conclusion: "The car has displayed characteristics that would seem to recommend it for further development." The original idea had been for Germany's carmakers to get together and produce the 'People's Car' jointly, but on July 4, 1936 the government decided to build

von den deutschen Automobilherstellern gemeinsam bauen zu lassen, fiel am 4. Juli 1936 die Entscheidung für den Bau des Volkswagenwerks. Am 28. Mai 1937 formierte sich die »Gesellschaft zur Vorbereitung des Deutschen Volkswagens mbH«, kurz Gezuvor. Als einer der drei Geschäftsführer erhielt Ferdinand Porsche den offiziellen Auftrag für Technik und Planung der zukünftigen Produktionsstätte. Um sich einen Überblick über den Stand der Massen-Produktionsverfahren zu verschaffen, besuchten Ferdinand und Ferry Porsche im Juni 1937 die Vereinigten Staaten von Amerika. In Detroit studierten sie modernste Herstellungsverfahren und versuchten, Fachleute aus der amerikanischen Industrie für das Volkswagen-Projekt zu gewinnen.

1938: PORSCHE IN ZUFFENHAUSEN

Um die provisorischen Zustände in der Garage der Porsche Villa zu beenden und die Bereiche Versuch und Konstruktion zusammenzuführen, ließ Ferdinand Porsche 1938 im Stadtteil Zuffenhausen ein neues Werk errichten. Das an der Spitalwaldstraße 2 gelegene Grundstück hatte Ferry Porsche bereits im Mai 1937 erworben und somit zugleich den Ort des heu-

RECHTE SEITE: Zeichnung Typ 114 vom 16. September 1938
RIGHT PAGE: *Drawing of the Type 114, September 16, 1938*

Volkswagen-Prototyp vor dem Haupteingang des Porsche-Werk 1 in Stuttgart-Zuffenhausen (1940)
The Volkswagen prototype in front of the main entrance to Porsche's #1 plant in Stuttgart-Zuffenhausen (1940)

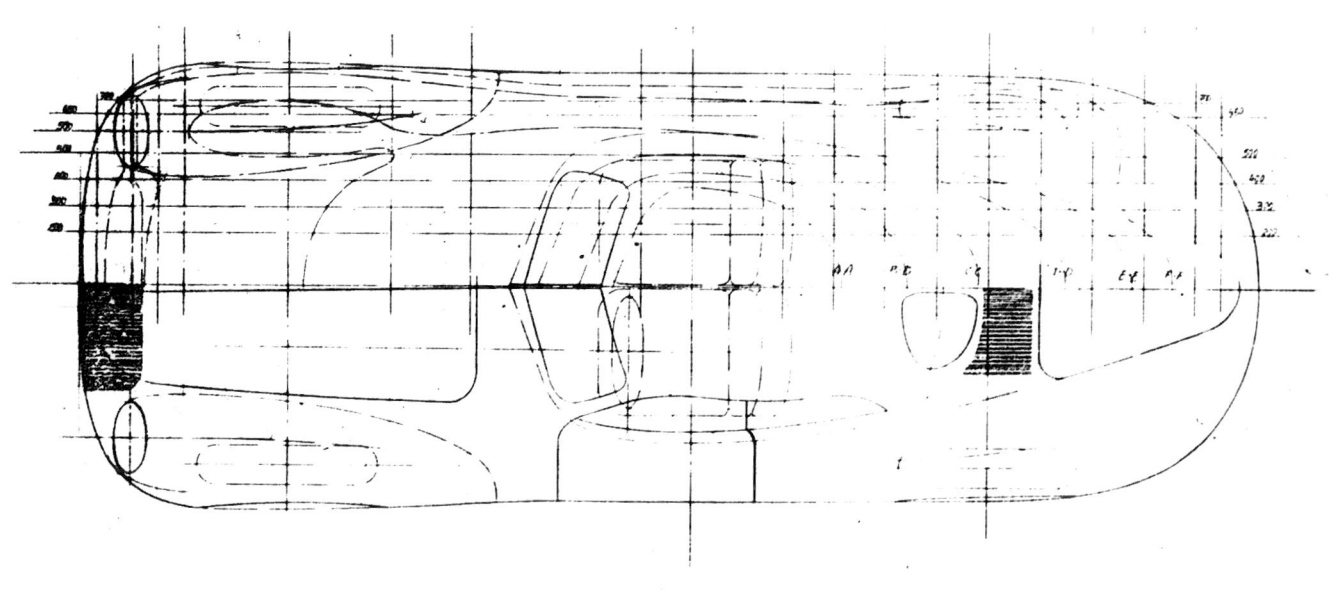

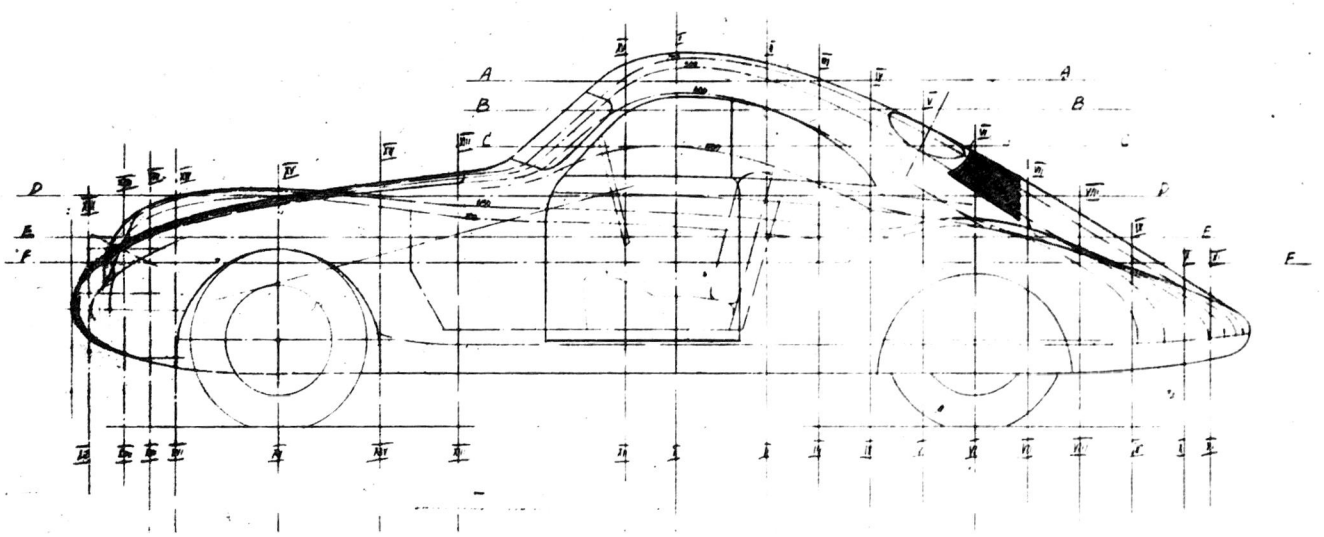

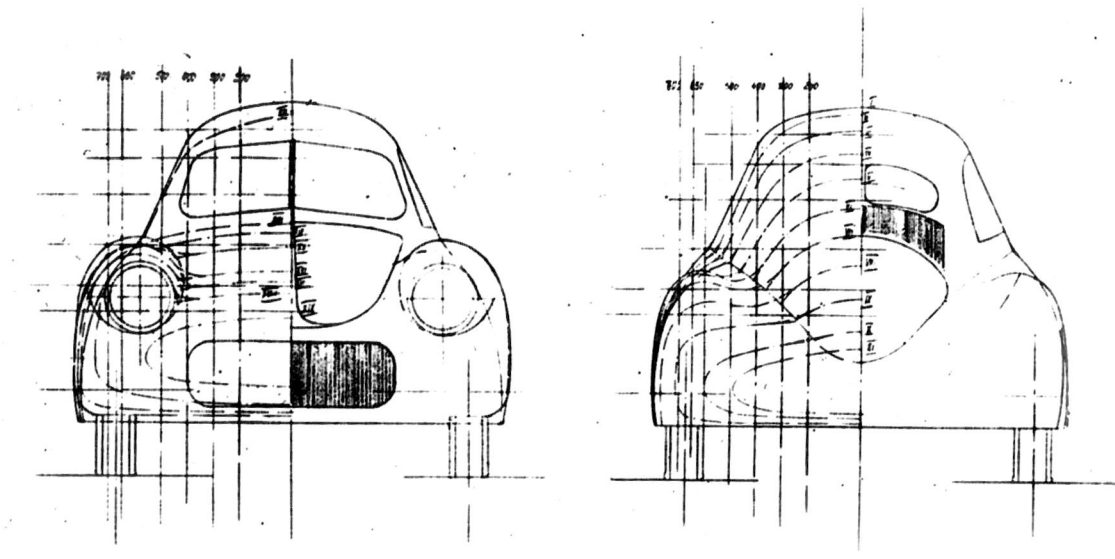

tigen Porsche-Stammwerks bestimmt. Im Juni 1938 entstand in Zuffenhausen die Nullserie des späteren VW-Käfers, für dessen Entwicklung Ferry Porsche immer mehr Verantwortung übernahm. »Am liebsten bin ich den VW auf einem nackten Fahrgestell, mit festgeschraubtem Sitz und ohne Karosserie gefahren – man sah, wie die Räder sich drehten und man spürte den Fahrtwind. Aus dieser Zeit stammt meine Vorliebe für klar definierte Ecken und Kotflügel – deshalb hatte unser Typ 60 auch vier richtige Kotflügel.«

Die Entwicklung des Volkswagens bedeutete nicht, dass Vater und Sohn Porsche ihre Lieblingsprojekte, die Konstruktion und Entwicklung von Renn- und Sportwagen, aufgegeben hatten. Ende der dreißiger Jahre entstand erstmals die Idee, eine eigene Fahrzeug-Produktion aufzubauen. Ferry Porsche dachte schon 1938 über einen kleinen Sportwagen auf Volkswagen-Basis nach. Versuchsweise rüstete er sein VW39-Cabriolet mit einem Kompressor-Motor aus. Doch die politische Führung verbot es bald darauf, Versuchsmotoren vom Volkswagenwerk zu erwerben, da diese für Rüstungszwecke benötigt wurden. Daraufhin entwickelten Vater und Sohn einen eigenen 1,5-Liter-Sportwagen nach dem Grundkonzept

Portrait (um 1939)
Portrait (about 1939)

RECHTE SEITE: Der Porsche Typ 64 »Berlin-Rom-Wagen« im Jahr 1939
RIGHT PAGE: *The Porsche Type 64 'Berlin-Rome' car, 1939*

a new production facility, the Volkswagenwerk. On May 28, 1937 a company was formed to prepare the way for the car's production, the 'Gesellschaft zur Vorbereitung des Deutschen Volkswagens mbH', or 'Gezuvor' for short. As one of its three general managers, Ferdinand Porsche was officially responsible for technical and planning work on the new production plant. To bring their ideas on the mass production of vehicles up to date, Ferdinand and Ferry Porsche visited the United States of America in June 1937. In Detroit, they were able to examine the very latest production processes, and also attempted to persuade experts from the American automobile industry to join the Volkswagen project.

1938: PORSCHE IN ZUFFENHAUSEN

To put an end to the makeshift situation in the garage of the Porsche villa and enable the experimental and design departments to work under one roof, Ferdinand Porsche had a new plant built in 1938. The site, at Spitalwaldstrasse 2 in the Stuttgart suburb of Zuffenhausen, had been acquired in May 1937 and is still at the center of the present-day Porsche

des Auto Union-Rennwagens. Das Typ 114 genannte Fahrzeug sollte einen Mittelmotor mit zwei oben liegenden Nockenwellen und halbkugelförmigen Brennräumen erhalten. Das Getriebe wurde hinter der Hinterachse positioniert, vorn sollten drei Personen nebeneinander sitzen, der Fahrer mit Lenkrad in der Mitte oder links. »Dieses Projekt habe ich mit mehr Engagement als mein Vater und unsere Leitenden Mitarbeiter verfolgt. Ich war nämlich überzeugt, dass uns ein großer Nachkriegsmarkt erwartete«, so Ferry Porsche später.

Parallel zu diesem Projekt entstand im Auftrag des Volkswagenwerks ein Wettbewerbsfahrzeug auf der Basis des Volkswagen Typ 60. Die Rennwagenkonstruktion sollte als Werbemaßnahme für den so genannten »KdF-Wagen« bei einem geplanten Langstreckenrennen von Berlin nach Rom starten. Unter der internen Bezeichnung Typ 64 beziehungsweise Typ 60K10 entwickelten und bauten die Porsche-Ingenieure im Frühjahr 1939 drei Rennsport-Coupés für die im September des Jahres geplante »Non-Stop Geschwindigkeitsprüfung«. Da ein großer Teil des über 1.500 Kilometer dauernden Rennens auf den neuen

Während einer Versuchsfahrt mit einem VW38 (1938)
During a test run of with a VW 38, 1939

production plant. In June 1938 the Zuffenhausen plant was used to build a pre-production run of the new Volkswagen (the car that was to go down in automobile history later as the 'Beetle'). Ferry Porsche took on increasing responsibility for its development. "What I enjoyed driving most of all was the bare platform of the VW, with a seat screwed to it but otherwise no body. You could see the wheels turning and feel the wind rushing past. This is how I developed a preference for clearly defined corners and fenders, and perhaps explains why our Type 60 had four 'proper' fenders."

Development work on the Volkswagen didn't lead to the Porsches, father and son, abandoning their favorite racing and sports car projects. It was toward the end of the 1930s that the idea of producing their own car was first mooted. By 1938 Ferry Porsche was already making plans for a small Volkswagen-based sports car. As an experiment, he installed a supercharged engine in his VW39 convertible. Unfortunately the government authorities soon forbade him to buy experimental engines from Volkswagen, claiming that they were all needed for military purposes. The Porsches responded by designing their own 1.5-liter sports car, based

deutschen Autobahnen stattfinden sollte, wurde der Aerodynamik des Fahrzeugs besondere Aufmerksamkeit geschenkt. Mit einer windschnittigen Stromlinienkarosserie aus Aluminium, verkleideten Radkästen sowie einem modifizierten VW-Vierzylinder-Boxermotor mit zunächst 33 PS erreichte der »Berlin-Rom-Wagen« eine Spitzengeschwindigkeit von bis zu 145 Stundenkilometern. Als der Kriegsbeginn die Austragung des Rennens verhinderte, benutze das Konstruktionsbüro Porsche die fertig gestellten Typ 64 als schnelle Reisesportwagen, mit denen auf Fahrten von Stuttgart nach Berlin Durchschnittsgeschwindigkeiten von 130 km/h erzielt wurden.

Mit jedem dieser Projekte trat Ferry Porsche ein wenig mehr aus dem Schatten seines Vaters Ferdinand. Obwohl er die größte Hochachtung vor den Fähigkeiten seines Vaters hatte, waren »wir Porsches auf technischem Gebiet keineswegs immer derselben Meinung«, konstatierte er in seinen Erinnerungen. »Wenn ich eine Meinung, die im Gegensatz zu seiner stand, in Gegenwart anderer äußerte, wurde er böse. Er fürchtete, glaube ich, sein Gesicht zu verlieren. Ergaben sich jedoch solche Widersprüche in unseren

Ferry Porsche mit seinen Söhnen Hans-Peter (links) und Wolfgang (rechts) auf dem Schüttgut 1943
Ferry Porsche with his sons Hans-Peter (left) and Wolfgang (right) at the 'Schüttgut' 1943

fundamentally on the Auto Union Grand Prix racer and known internally as the Type 114. The centrally located mid-engine was to have two overhead camshafts and hemispherical combustion chambers. The gearbox was behind the rear axle, and the three occupants were to sit in a single row, with the driver and steering wheel either on the left or in the center. Ferry Porsche, looking back later: "I tackled this project with more enthusiasm than my father or our senior staff, because I was confident that there would be a big market for it after the war."

In parallel with this project, the Volkswagenwerk commissioned a competition car for the planned Berlin - Rome road race, based on the Volkswagen Type 60. This was a publicity project to advertise the people's car, which had in the meantime acquired the name "KdF-Wagen". With the internal designations Type 64 and Type 60K10, Porsche's engineers developed and built three sports racing coupes for the "non-stop high-speed test" scheduled to take place in September. Since much of the 1,500-kilometer route was to make use of Germany's new 'autobahn' highways, special attention was paid to the

Ferry Porsche am Steuer eines Typ 128 bei Schwimmversuchen im Löschwasserteich des Porsche-Werks in Stuttgart-Zuffenhausen
Ferry Porsche at the wheel of a Type 128 amphibious vehicle, during flotation tests on the Porsche plant's firefighting pond in Stuttgart-Zuffenhausen

RECHTE SEITE:
Erprobung der Windkraftanlage Typ 135 auf dem Porsche-Werksgelände: (V.l.n.r.) Herr Tschentscher, Bodo Lafferentz, Josef Mickl, Ferdinand und Ferry Porsche

Ganz rechts: Das Porsche-Werk in Gmünd, für das Ferry Porsche 1946 die Gesamtverantwortung übernahm

RIGHT PAGE:
Testing the Type 135 wind power unit at the Porsche plant: (from left to right) Mr. Tschentscher, Bodo Lafferentz, Josef Mickl, Ferdinand Porsche and Ferry Porsche

On the right: In 1946 Ferry Porsche took over entire responsibility for the Porsche plant in Gmünd, Austria

Ansichten, wenn wir beide allein waren, etwa auf einer längeren Autofahrt, dann war er viel zugänglicher und hörte sich geduldig meine Meinung an. Vater war eine sehr autoritäre Persönlichkeit.«

1944: PORSCHE IN GMÜND

Nach dem Ausbruch des Zweiten Weltkrieges beschäftigten sich die Porsche-Ingenieure vorwiegend mit der Entwicklung militärischer Fahrzeuge. Neben dem Typ 81 »VW-Kastenwagen« entwickelte das seit Ende 1937 als Porsche KG firmierende Unternehmen den Typ 62 »KdF-Gelände-Fahrzeug«, den als »VW-Kübelwagen« bekannt gewordenen Typ 82 sowie den mit Allradantrieb ausgestatteten Typ 87 und den Typ 166 »VW-Schwimmwagen«. Vom Heereswaffenamt erhielt das Porsche-Konstruktionsbüro zudem Ende 1939 den Entwicklungsauftrag für einen mittelschweren Kampfpanzer, dessen Konstruktion jedoch vorzeitig eingestellt wurde, da schwerere Panzertypen benötigt wurden.
Ab 1943 wurde das Leben der Porsche-Mitarbeiter immer stärker durch die Luftangriffe auf Stuttgart geprägt. Auf Drängen des Rüstungskommandos der

LINKE SEITE: Stand der Dr.Ing.h.c.F.Porsche KG auf der Industrie- und Gewerbeausstellung in Klagenfurt vom 8. bis zum 22. August 1946
LEFT PAGE: The Dr.Ing. h.c.F. Porsche KG stand at the Industry and Trade Exhibition in Klagenfurt, August 8–22, 1946

Dorothea und Ferry Porsche in Österreich (1946)
Dorothea and Ferry Porsche in Austria, 1946

car's aerodynamics. The streamlined aluminum body had closed-in wheel arches; the cars were powered by modified VW flat-four engines that developed 33 horsepower initially. In this form the 'Berlin-Rome' car could reach a top speed of 145 kilometers an hour. When war broke out and the race was canceled, the Porsche design office used the completed Type 64 cars as fast sport tourers capable of completing the run from Stuttgart to Berlin at an average speed of 130 kph.

With each of these projects, Ferry Porsche moved steadily out from under his father's wing, so to speak. Although he had the greatest possible admiration for his father's skills, he recalls in his memoirs that "we Porsches were certainly not always of the same opinion on technical matters. Father would be annoyed if I expressed an opinion that differed from his in the presence of other people. I imagine that he felt he was losing face in such situations. If our views conflicted when we were alone on a long car journey, on the other hand, he was much more willing to listen patiently to what I had to say. My father was an extremely authoritarian personality."

Unter der Regie von Ferry Porsche entstand 1948 der Typ 360 »Cisitalia«. 3.v.l.: Karl Abarth; 4.v.l.: Pierro Dusio; 3.v.r. Rudolf Hruska; 4.v.r. Tazio Nuvolari

Ferry Porsche supervised development of the Type 360 'Cisitalia' in 1948. Third from left: Karl Abarth; fourth from left: Piero Dusio; third from right: Rudolf Hruska; fourth from right: Tazio Nuvolari

Wehrmacht musste die Porsche KG schließlich im Herbst 1944 das Konstruktionsbüro von Stuttgart nach Gmünd in Kärnten auslagern. Auf dem Gelände der »W. Meineke Holzgroßindustrie Berlin-Gmünd« wurden Behelfswerkstätten aufgebaut, während das Materiallager auf dem Areal einer Fliegerschule im nahe gelegenen Zell am See untergebracht wurde. In eine Vielzahl von Arbeitsstätten und Unterkünften aufgesplittert, erhielt das neue Porsche-Werk von den Mitarbeitern den spöttischen Beinamen »Vereinigte Hüttenwerke«.

Nach Kriegsende wurden die Zuffenhausener Produktionsanlagen der Dr. Ing. h.c. F. Porsche KG zunächst vom französischen Militär genutzt. Im August 1945 übernahm eine amerikanische Einheit das zur Reparaturwerkstätte für Lastwagen umfunktionierte Werk 1. Das Porsche-Werk in Gmünd erhielt unterdessen eine provisorische Bewilligung zur Wiederaufnahme der Arbeit. Den rund 140 Porsche-Mitarbeitern wurde es gestattet, »Entwürfe von Motor-Traktoren, Gaserzeugern und anderen zivilen Einrichtungen« durchzuführen sowie »Motorfahrzeuge und landwirtschaftliche Maschinen« zu reparieren.

1944: PORSCHE IN GMÜND

After the Second World War began Porsche's engineers were mainly concerned with the development of military vehicles. In addition to the Type 81, the VW 'Kastenwagen', the company – which had changed its commercial status at the end of 1937 to 'Porsche KG' – developed the Type 62 'KdF Offroad Vehicle', known generally as the 'Kübelwagen' (literally 'Bucket Car'), the Type 87 with all-wheel drive and the Type 166, the 'VW Amphibian'. At the end of 1939 the German Army Procurement Office commissioned the Porsche Design Office to develop a medium-weight battle tank. Work on it was prematurely halted, however, when the army decided that heavy tanks were needed more urgently.

From 1943 on, Porsche's workforce had to contend increasingly with bombing raids on Stuttgart. In the fall of 1944, the Armaments Command insisted that the design office should move from Stuttgart to the town of Gmünd in Carinthia, Austria. Part of the site occupied by the timber wholesaler 'W. Meineke Holzgrossindustrie Berlin-Gmünd' was used to build temporary workshops, and a material store was set

Der Kompressor-Zwölfzylindermotor des Typ 360 leistete 385 PS bei nur 1,5 Liter Hubraum
The supercharged twelve-cylinder engine for the Type 360 had a power output of 385 bhp from a displacement of only 1.5 liters

In dieser schwierigen Situation folgte Ferdinand Porsche Mitte November 1945 der Einladung einer französischen Kommission nach Baden-Baden, um eine Fortsetzung des Volkswagen-Projekts in Frankreich zu besprechen. Bevor es jedoch einen Monat darauf bei einem weiteren Treffen zu einem Vertragsabschluss kam, wurde Ferdinand Porsche zusammen mit seinem Sohn Ferry und Schwiegersohn Anton Piëch vom französischen Geheimdienst in Baden-Baden verhaftet. Während Ferry Porsche im

up at a flying school in nearby Zell am See. The new Porsche design office was spread among so many workshops and buildings that the staff referred to it ironically as the 'United Hut Corporation'.

When the war came to an end, the production facilities of 'Dr. Ing. h.c. F. Porsche KG' in Zuffenhausen were used initially by the French Army. In August 1945 an American military unit took over the #1 plant and used it for truck repairs. The Porsche plant in Gmünd, on the other hand, was granted a provisional permit to start operating again. A workforce of about 140 was allowed to "design motor tractors, gas generators and other equipment for civil use", and also to repair motor vehicles and agricultural machinery. In this difficult situation, Ferdinand Porsche complied with an invitation from a French inquiry commission to come to Baden-Baden, in the southwest of Germany, in November 1945, to discuss continuation of the 'People's Car' project in France. A month later, there was to be a further meeting with the aim of concluding an agreement, but before this Ferdinand Porsche, his son Ferry and his son-in-law Anton Piëch were arrested in Baden-Baden by the French Secret

März 1946 wieder aus dem Gefängnis frei kam, hielt man den Senior trotz einer schweren Erkrankung weiter fest und brachte ihn nach Paris und Dijon.

»Nach dem Krieg wurde es für mich ernst, denn nun kam es alleine auf meine Initiative an«, erinnerte sich Ferry Porsche. Er nahm neue Projekte in Angriff, die auf die veränderten Verhältnisse zugeschnitten waren: Man benutzte die Werkzeugmaschinen in Gmünd dazu, Zubehörteile für Traktoren, Handkarren oder Seilwinden zu bauen. Das große Ziel war es aber, wieder Autos zu konstruieren. Auf Vermittlung seiner Geschäftsfreunde Karl Abarth und Rudolf Hruska unterzeichnete Ferry Porsche am 15. Dezember 1946 einen Vertrag mit dem Turiner Industriellen Piero Dusio über umfangreiche Entwicklungsaufträge für dessen Firma Cisitalia. Neben einem kleinen Traktor und einer Wasserturbine begann die Gmünder Porsche KG mit der Auftragskonstruktion eines Typ 360 Grand-Prix-Rennwagens sowie des zweisitzigen Typ 370 Mittelmotor-Sportwagens.
Unter Hochdruck arbeiteten die Konstrukteure der Gmünder Porsche KG am Grand-Prix-Rennwagen Typ 360. Das erstmals komplett unter der Regie von

Service. Ferry Porsche was released from prison in March 1946, but his father, although seriously ill, remained under arrest and was subsequently moved to Paris and Dijon.

Ferry Porsche, looking back later: "My situation after the war was serious, since everything depended on my own initiative." He began new projects that took the changed circumstances into account: the machine tools in were used to make spare parts for tractors, handcarts and rope winches. The long-term objective, however, was to design and build cars. Ferry Porsche's friends Karl Abarth and Rudolf Hruska put him in touch with the Turin-based industrialist Piero Dusio, and on December 15, 1946 a contract was signed for extensive development work to be carried out on behalf of Dusio's Cisitalia company. In addition to a small tractor and a water turbine, Porsche KG began from its Gmünd location to design the Type 360 Grand Prix racing car and the Type 370 mid-engine two-seat sports car for this client.
The designers in Gmünd tackled the Type 360 Grand Prix racing car project as a matter of urgency.
This was the first car to be developed with Ferry

Ferry Porsche entwickelte Fahrzeug war seiner Zeit weit voraus: Als Antrieb sahen die Porsche-Ingenieure einen 1,5-Liter-Zwölfzylindermotor mit Kompressoraufladung vor, die Kraftübertragung erfolgte über einen zuschaltbaren Vierrad-Antrieb. Obwohl der Auftrag aufgrund von Kapitalmangel des italienischen Auftraggebers nicht über ein Versuchsstadium hinaus geführt werden konnte, war das Projekt von entscheidender Bedeutung. Durch das Honorar konnte Ferry Porsche eine Kaution für seinen inhaftierten Vater in Höhe von einer Million französischen Franc aufbringen, der daraufhin am 1. August 1947 entlassen wurde. Die in Frankreich wegen angeblicher Kriegsvergehen eingeleitete Untersuchung gegen Professor Porsche gelangte nicht zur formellen Anklageerhebung und wurde wenig später endgültig eingestellt. Als Ferdinand Porsche kurz darauf in Gmünd die Zeichnungen des Typ 360 betrachtete, gab er der Arbeit seines Sohnes sein bestmögliches Urteil: »Keine Schraube hätte ich anders gemacht.«

Typ 360 »Cisitalia« auf dem Hof des Porsche-Werks in Stuttgart-Zuffenhausen (1961)
The Type 360 'Cisitalia' in the yard of the Porsche plant in Stuttgart-Zuffenhausen (1961)

Porsche entirely responsible, and the design was well ahead of its time. Porsche's engineers proposed a supercharged 1.5-liter twelve-cylinder engine and a transmission with driver-engaged four-wheel drive. Although the Italian client ran short of funds and the project never progressed beyond the experimental stage, it was of decisive importance for Porsche. The fee that the company charged enabled Ferry Porsche to meet the French demand for a million francs as bail for his father, who was thereupon released from prison on August 1, 1947. The investigation alleging Professor Porsche's involvement in war crimes that was begun in France never resulted in a formal arraignment and was closed after a short time. In due course, when Ferdinand Porsche returned to Gmünd and looked through his son's drawings for the Type 360, he commented: "I wouldn't wish to change a single bolt or nut!" No son could have wished for a better verdict on his work.

92 | 93 **LEHRJAHRE IM KONSTRUKTIONSBÜRO** | EARLY YEARS IN THE DESIGN OFFICE

DER ERSTE PORSCHE-SPORTWAGEN
THE FIRST PORSCHE SPORTS CAR

DER ERSTE PORSCHE-SPORTWAGEN
THE FIRST PORSCHE SPORTS CAR

Im Frühjahr 1947 formulierte Ferry Porsche seine ersten Überlegungen zum Bau eines auf Teilen des Volkswagen basierenden Sportwagens, der, zunächst als »VW-Sport« bezeichnet, die Konstruktionsnummer 356 erhielt. Der Porsche Junior-Chef hatte die Vision, »einen Sportwagen zu bauen, wie er mir selbst gefiel«. Anlässlich seines 75. Geburtstages erzählte Ferry Porsche über die Anfänge des Porsche 356: »Die Anregung kam, das kann man ruhig zugeben, durch Cisitalia. Diese Firma baute damals einen kleinen Sportwagen mit Fiat-Motor. Da sagte ich mir: Warum sollten wir nicht das Gleiche mit VW-Teilen tun können? Ähnliches hatten wir schon vor dem Krieg mit dem Berlin-Rom-Wagen getan. [...] Uns schwebte ein kleines, wendiges, leichtes Fahrzeug vor, das die Leistungen eines großen, leistungsstarken Wagens

Ferdinand Porsche (rechts), Sohn Ferry (mitte) und Mitarbeiter Erwin Komenda mit dem Typ 356 »Nr.1« in Gmünd (1948)
Ferdinand Porsche (right), his son Ferry (center) and Erwin Komenda from the Porsche staff with the Type 356 'No. 1' in Gmünd (1948)

In the spring of 1947 Ferry Porsche formulated his ideas for a sports car based on the use of Volkswagen components. The project was known initially as 'VW Sport' and given the design number 356. Porsche's junior chief executive had a vision of "building a sports car the way I would like it!" At his 75th birthday celebrations, Ferry Porsche spoke about the early days of the Porsche 356 concept: "It's only fair to say that the suggestion came from the Cisitalia company, which was building a small sports car with a Fiat engine at that time. I said to myself: 'Why shouldn't we do the same with VW parts?' After all, we had done something similar before the war when we built the Berlin-Rome car [...] What we envisaged was a small, agile, lightweight car that would perform better than many a bigger car with a more powerful engine. We

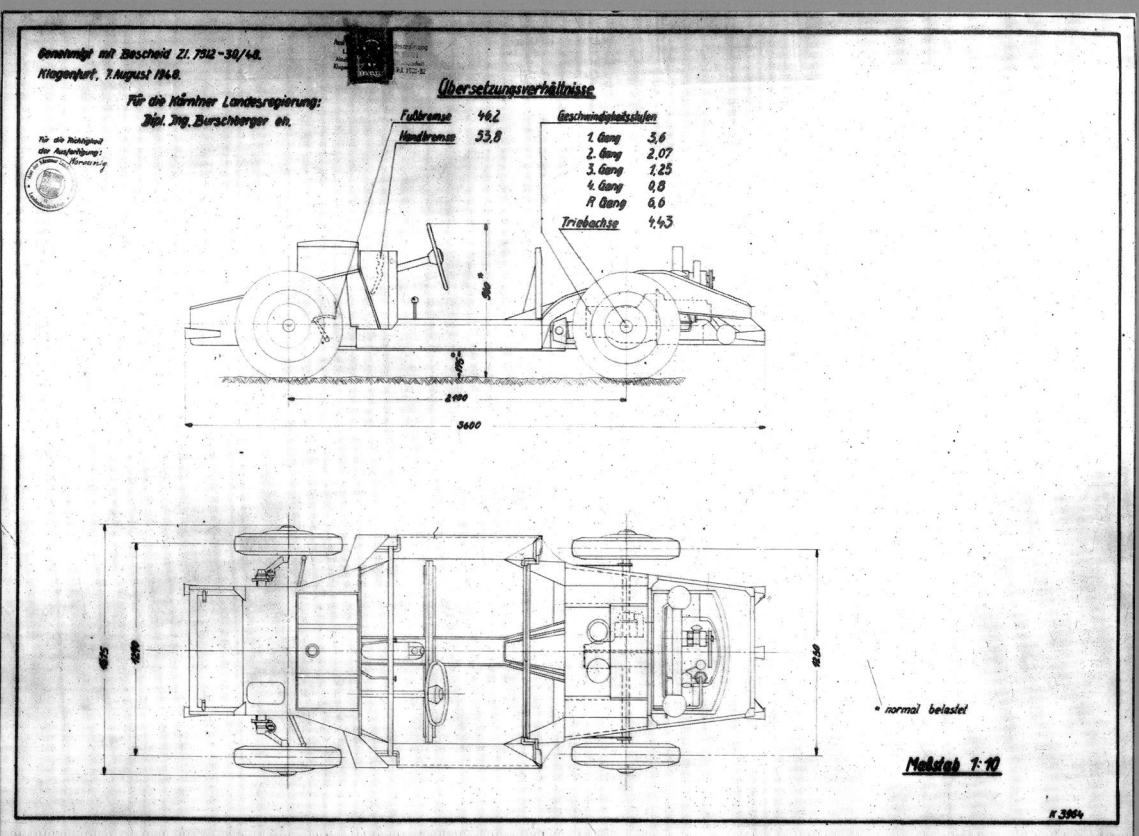

LINKE SEITE: Der erste Sportwagen mit dem Namen Porsche: Typ 356 »Nr.1« (1948)
LEFT PAGE: *The first sports car to carry the Porsche name: the Type 356 'No. 1' (1948)*

Konstruktionszeichnung des Chassis Typ 356/2 (1948)
Right page: Design drawing for the Type 356/2 chassis (1948)

1948: Produktion der
Porsche 356/2 in Gmünd,
Kärnten
*1948: Production of the
Porsche 356/2 in Gmünd,
Austria*

übertreffen sollte. Und so planten wir, 50 solcher Automobile in Gmünd zu bauen.« Bei der Technik orientierte sich Ferry Porsche an seinen Erfahrungen aus der »VW-Käfer«-Entwicklung: »Das Konzept formte sich schon in meinem Kopf, während ich den Volkswagen zur Serienreife brachte. Das Geheimnis ist, dass bedingt durch die Motorlage im Heck und die Sitze im Schwerpunkt des Wagens, immer dafür gesorgt ist, die größte Leistung des Motors bei Zweiradantrieb auf den Boden zu bringen.«

Die Ingenieure um Ferry Porsche waren fasziniert von der Sportwagen-Idee und so entstand schon im Februar 1948 ein fahrbereites Fahrgestell, für das ein schnittiger Roadster-Aufbau aus Aluminium angefertigt wurde. Der Vierzylinder-Boxermotor stammte ebenso wie Getriebe, Radaufhängung, Federung und Lenkung aus dem Volkswagen. Der 35 PS starke Mittelmotor-Roadster erreichte bei nur 585 Kilogramm eine Höchstgeschwindigkeit von 135 km/h.
Den offiziellen Segen der Behörden erhielt der mit der Fahrgestellnummer 356-001 versehene Mittelmotor-Sportwagen am 8. Juni 1948 durch eine Einzelgenehmigung der Kärntner Landesregierung.

Ferry Porsche auf dem Schüttgut (1949)
Ferry Porsche at the 'Schüttgut', 1949

decided to build a batch of fifty cars of this type in Gmünd." For the technical details, Ferry Porsche drew on his experience with the VW Beetle: "The concept took shape in my mind while I was bringing the Volkswagen to production maturity. The secret is that if the engine is at the back and the occupants sit close to the car's center of gravity, you can always be certain that however powerful the engine, it will transmit its power to the road even with two-wheel drive."

Ferry Porsche's engineering colleagues were fascinated by the idea of developing a sports car. By February 1948 a roadgoing chassis had been built and clothed with a streamlined aluminum roadster body. The flat-four engine, gearbox, suspension, springs and steering all came from the Volkswagen. This mid-engine roadster, with its power output of 35 hp, weighed only 585 kilograms and could reach a top speed of 135 kph.
Official approval of the new mid-engine sports car with chassis number 356-001 was granted on June 8, 1948 by the Carinthian regional government in the form of a special permit. Today we can only marvel

Das wirtschaftliche Risiko dieser Unternehmung lässt heute erstaunen. Ganz Europa befand sich im Wiederaufbau, die Massenmotorisierung lief gerade erst an und es waren vor allem praktische und preiswerte Fahrzeuge, die ihren Weg auf den Markt fanden. In dieser Situation realisierte Ferry Porsche seinen Traum vom eigenen Sportwagen – und stellte fest, dass andere Automobilliebhaber diesen Traum mit ihm teilten. Noch in der zweiten Jahreshälfte 1948 begann in Gmünd die Fertigung der ersten serienmäßigen Coupés und Cabriolets vom Typ 356/2. Wie der Porsche 356 »Nr. 1« erhielten auch die 356/2 eine von Erwin Komenda, dem Leiter der Porsche-Karosserieentwicklung, gestaltete Aluminiumkarosserie. Doch im Gegensatz zum Mittelmotor-Prototyp »Nr.1« wurde der Boxermotor des Typ 356/2 im Heck positioniert, um hinter den Vordersitzen einen Gepäckraum zu ermöglichen. In Zürich fand sich mit Rupprecht von Senger ein Investor, der Geld für eine Kleinserie vorschoss und dafür einen Vertrag als Importeur für die Schweiz erhielt. Ein Glücksfall, denn über diesen Kontakt erhielt Porsche auch dringend benötigte VW-Teile und Karosseriebleche.
Dass Ferry Porsche neben einem guten technischen

at the risk that the project involved. The whole of Europe was recovering from the hardships and destruction of the war years, mass motorization was in its infancy and what the population needed were surely the practical, reasonably priced vehicles that were beginning to find their way onto the market. This was the situation in which Ferry Porsche decided to make his sports-car dream come true – and soon discovered that other car enthusiasts shared the same vision. The first production coupes and convertibles were built in Gmünd in the second half of 1948. Like 'Number One', the very first Porsche 356, the 356/2 had an aluminum body styled by Erwin Komenda, Porsche's Head of Body Development. In contrast to the first mid-engine prototype, however, the Type 356/2 had its flat-four engine at the rear in order to create luggage space behind the seats. Rupprecht von Senger, an investor from Zürich, advanced the money for the first batch of cars, and was invited to act as importer for Switzerland. This was a fortunate decision, since he had contacts with VW and was able to obtain the parts and body stampings that Porsche urgently needed.
That Ferry Porsche's instinctive talent for technical

LINKE SEITE: Ferry Porsche (2.v.l.) am Chassis des Typ 356/2 in Gmünd, Kärnten (1948)

LEFT PAGE: Ferry Porsche (second from left) with the Type 356/2 chassis in Gmünd, Carinthia (1948)

Ferry Porsche im Typ 356/2 »Gmünd-Coupé«. Kinder (v.l.): Ferdinand Piëch, Edwin Kaes, Michel Piëch (1948)
Ferry Porsche in the Type 356/2 'Gmünd Coupé'. The children (from left) are Ferdinand Piëch, Edwin Kaes and Michel Piëch (1948)

Gespür auch unternehmerische Weitsicht besaß, beweist der am 17. September 1948 mit dem Volkswagenwerk-Geschäftsführer Heinrich Nordhoff ausgehandelte Vertrag über die Zulieferung von VW-Teilen sowie die Nutzung des VW-Vertriebsnetzes. Ferry Porsche erreichte, dass VW für jeden gebauten Käfer eine Lizenzgebühr an Porsche zahlte, da dieser schließlich vor dem Krieg von Porsche entwickelt worden war. Außerdem wurde mit der »Porsche-Salzburg Ges. m.b.H.« ein zentrales Büro zur Steuerung von Import, Vertrieb und Kundendienst für den Volkswagen in Österreich eingerichtet. Diese Vereinbarungen mit dem großen Volkswagenwerk bedeuteten Sicherheit für das junge Unternehmen Porsche – besonders in finanzieller Hinsicht. Die Grundlage für den Ausbau der Porsche KG als Sportwagenhersteller war geschaffen.

1950: DIE RÜCKKEHR NACH STUTTGART

Mit zunehmendem Erfolg des Porsche 356 wurde offensichtlich, dass das provisorische Werk in Gmünd nicht ausreichte, um den Sportwagenbau weiter forcieren zu können. Zu mangelhaft war die technische Ausrüstung der österreichischen Fertigungsstätte,

Mit dem Volkswagenwerk-Geschäftsführer Heinrich Nordhoff (1952)
With Heinrich Nordhoff, General Manager of the Volkswagenwerk (1952)

matters was accompanied by the accomplishments of a far-sighted businessman was proved by the contract that he concluded on September 17, 1948 with Heinrich Nordhoff, the Volkswagenwerk's General Manager. This assured Porsche of a supply of VW parts and access to the VW sales network. Ferry Porsche also persuaded the Wolfsburg-based company to pay a license fee for every 'Beetle' it built, since the pre-war design was in fact a Porsche development. 'Porsche-Salzburg Ges.m.b.H.' was set up for Volkswagen in Austria as a central office for imports, sales and customer service. This agreement with the very much larger Volkswagenwerk meant security for the young Porsche company – especially in financial terms – and created a sound basis for Porsche KG to develop as a sports-car manufacturer.

1950: RETURN TO STUTTGART

The Porsche 356's increasing success made it obvious that the provisional factory in Gmünd did not have the capacity to boost sports-car sales any further. Its production facilities were inadequate, and business conditions in this Alpine province of Austria

Ferry Porsche (3.v.r.) und Vater Ferdinand (4.v.r.) mit einer Rohkarosse des Typ 356 im Hof der Karosseriefabrik Reutter in der Augustenstraße 82 in Stuttgart-West (1950)
Ferry Porsche (third from right) with his father Ferdinand (fourth from right) and a Type 356 bodyshell, in the yard of the Reutter coachbuilding company on Augustenstrasse 82, Stuttgart-West (1950)

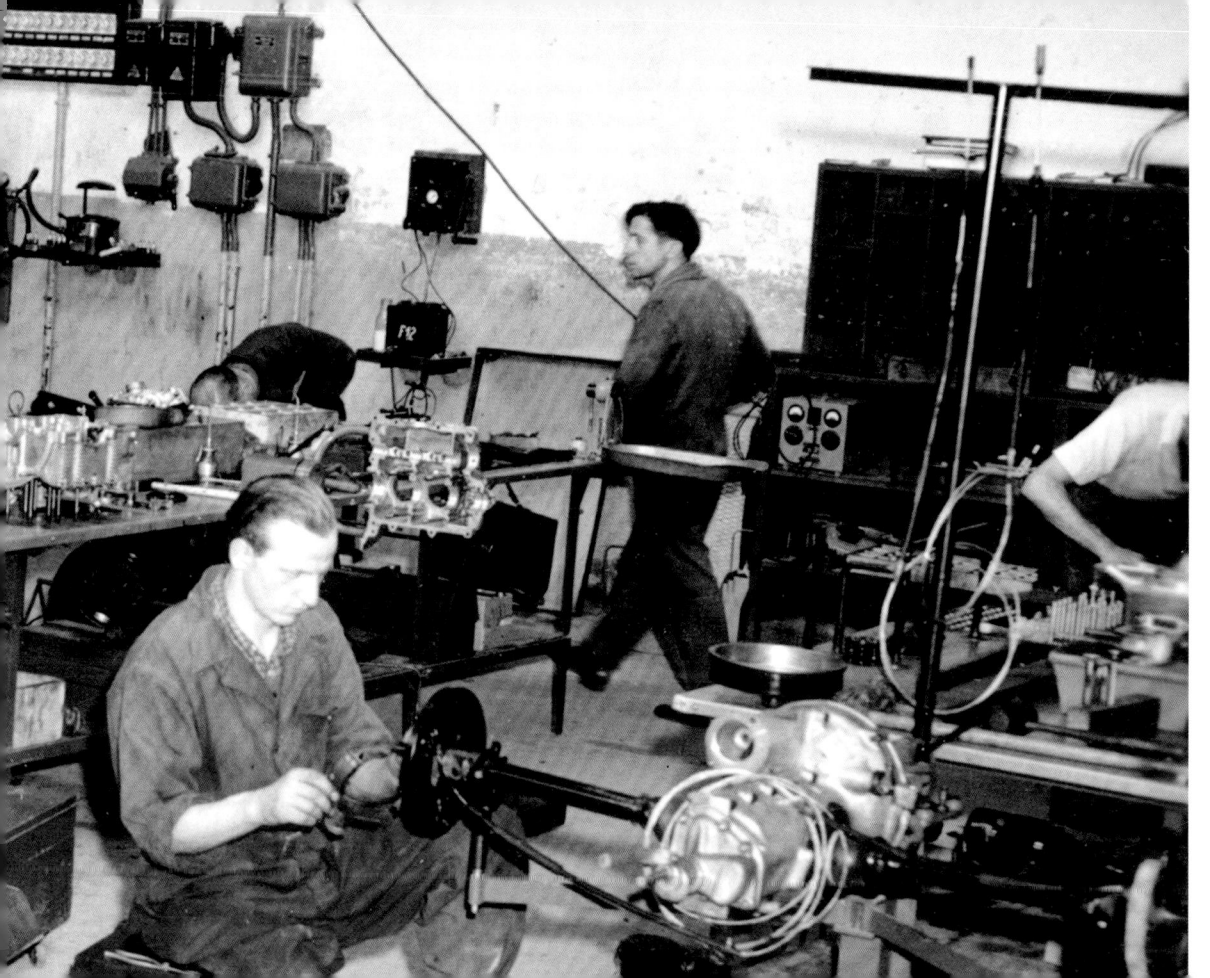

1950: Ferdinand Porsche (3. v. l.) auf einem Rundgang durch die Porsche-Werkstatt in Stuttgart-Zuffenhausen. Neben ihm (rechts) sein Sohn Ferry
1950: Ferdinand Porsche (third from left) on a tour of inspection of the Porsche workshop in Stuttgart-Zuffenhausen. On his left, his son Sohn Ferry

und als zu schwierig erwiesen sich die Wirtschaftsbedingungen des Alpenlandes. Dass die Zukunft des Unternehmens letztendlich im Sportwagenbau liegen würde, war zu diesem Zeitpunkt nicht entschieden. Zwar stimmten die ersten Verkaufserfolge des Typ 356/2 optimistisch, doch das Augenmerk des Senior-Chefs Ferdinand Porsche lag auf den Diesel-Schleppern und Wasserturbinen. Für den traditionellen Bereich der Auftragsentwicklung erwartete er mehr Erfolg als mit einer eigenen Fahrzeugfertigung. Ferry Porsche hingegen glaubte an die Kraft seiner Idee und wollte zumindest eine Serie von einigen hundert Fahrzeugen realisieren. Dazu waren die Verhältnisse in Gmünd jedoch zu provisorisch und so entschloss er sich, in die Autostadt Stuttgart zurückzukehren.

Da das ehemalige Porsche-Werk in der Zuffenhausener Spitalwaldstraße 2 noch immer von den Amerikanern genutzt wurde, entschied Ferry Porsche, vorläufig ein Büro und eine kleine Versuchswerkstatt in der Stuttgarter Porsche-Villa einzurichten. Die Umzugsvorbereitungen übernahm sein Schulfreund Albert Prinzing, der im November 1949 als Mitgeschäftsführer der Stuttgarter Porsche Konstruktionen GmbH eingesetzt

Vater und Sohn mit dem Typ 356 vor der Garage der Porsche-Villa (1950)
Father and son with the Type 356 in front of the garage at the Porsche villa (1950)

were difficult. At this time it was not yet clear whether the company's future lay in building sports cars. Initial sales of the Type 356/2 seemed to justify an optimistic attitude, but patriarch Ferdinand Porsche regarded the prospects for diesel tractors and water turbines as more promising. He felt that Porsche's traditional activities, involving the development of designs commissioned by other companies, were likely to yield better results than producing its own cars. Ferry Porsche on the other hand was convinced of the merits of his own idea, and wished to build at least a small batch of a few hundred cars. But for this, conditions in Gmünd were too temporary in nature, and he therefore resolved to move back to Stuttgart.

Since the Americans were still using the former Porsche plant at Spitalwaldstrasse 2 in Zuffenhausen, Ferry Porsche decided to begin operations with an office and a small experimental workshop in the Porsche villa in Stuttgart. Preparations for the move were in the hands of Albert Prinzing, who had been appointed joint general manager of Porsche Konstruktionen GmbH based in Stuttgart in 1949. At the end of the year Porsche GmbH rented a

Ferdinand Porsche (rechts) an seinem 75. Geburtstag am 3. September 1950. In der Mitte Ferry, links Rennlegende Rudolf Caracciola

Ferdinand Porsche (right) on his 75th birthday (September 3, 1950). Ferry Porsche is in the center, legendary racing driver Rudolf Caracciola at the right

wurde. Bei der in Stuttgart-Zuffenhausen ansässigen Karosseriewerke Reutter & Co. GmbH mietete die Porsche GmbH zum Jahresende eine 600 Quadratmeter große Halle an und vergab einen Fertigungsauftrag über 500 Karosserien an Reutter. »Weil Reutter mit Leichtmetall-Schweißen keine Erfahrung hatte, mussten wir das Coupé auf Stahl umstellen«, entschied Ferry Porsche. Im März 1950 wurde der erste Porsche 356 in Stuttgart gebaut. Während sich Ferry Porsche aus Zeitgründen zunehmend aus der Konstruktion zurückziehen musste und Managementaufgaben übernahm, entwickelte sich der 356 zu einem Bestseller. Die freien Kapazitäten von Reutter waren schnell erschöpft und mehrere andere Karosseriefabriken sprangen ein. Und Ferry Porsche gestand freimütig ein, dass er sich bei der gebauten Stückzahl deutlich verschätzt hatte: »Ich dachte, dass ich vielleicht 500 solcher Porsche-Autos verkaufen könnte und damit meine Kosten begleichen würde. Es stellte sich heraus, dass das eine kleine Unterschätzung war – und von diesem Zeitpunkt an hat es hier immer viel zu tun gegeben. [...] Wir hätten es uns nicht träumen lassen, dass wir schließlich mit dem Typ 356 auf eine Gesamtstückzahl von 78.000 kommen würden«, resümierte Ferry Porsche zufrieden.

600 square-meter building from the coachbuilders Reutter & Co. GmbH, also situated in Stuttgart-Zuffenhausen, and ordered 500 car bodies from that company. "But Reutter had no experience of welding light alloys, so we had to give the coupe a steel body," Ferry Porsche recalls. The first Porsche 356 built in Stuttgart was produced in March 1950. Ferry Porsche himself had to withdraw increasingly from the design side of the business and devote himself to management tasks. It was soon obvious that the 356 was going to be a best-seller. Spare capacity at Reutter was soon used up, and bodies had to be built by several other companies. Ferry Porsche readily admitted that he had misjudged demand for the new car. Satisfied, he commented later: "I thought we might perhaps sell 500 Porsche cars of this type, and at least recover my costs. As it happened, this was a slight underestimate – and from that time on, we had our hands full. [...] In our wildest imagination, we never expected to sell the Type 356 model 78,000 times!"

Building up exports at an early stage was an important success factor for the young Porsche

Ferry Porsche (um 1950)
Ferry Porsche (about 1950)

Das 500. Exemplar des Porsche 356 entstand bereits 1950
By 1950, five hundred Porsche 356 cars had already been built

Ein wichtiger Erfolgsfaktor für das junge Unternehmen Porsche war die frühe Exportorientierung. Auf dem Genfer Salon wurde der Porsche 356 als Coupé und Cabriolet 1949 erstmals der Öffentlichkeit präsentiert. Ein Jahr darauf ließ Ferry Porsche die ersten Wagen nach Amerika verschiffen, um sie über das Händlernetz des geschäftstüchtigen Automobilimporteurs Max Hoffman verkaufen zu lassen. Auf dem größten und wichtigsten Absatzmarkt der Welt eroberte der Porsche 356 die Herzen der Sportfahrer – und nicht zuletzt auch vieler Hollywoodstars – im Sturm. Mit Modellen wie dem 356 Speedster traf Ferry Porsche den Geschmack der amerikanischen Kunden, an die bereits 1955 die Hälfte der Jahresproduktion verkauft wurde. Neben dem Export war es auch Ferry Porsches Leidenschaft für den Motorsport, die zum Erfolgsgaranten der Marke wurde. Anstatt Reklame oder Werbung zu betreiben, sollten seine Sportwagen vor den Augen des Publikums durch Rennsiege für sich selbst sprechen: »Die extremen Beanspruchungen bei Rennen lassen bald die schwachen Stellen erkennen und regen damit den Techniker an, neue, bessere Wege zu suchen.« Das Resultat waren schnelle und zuverlässige Automobile, die auf der ganzen Welt von

Mit den Söhnen Ferdinand Alexander, Wolfgang und Hans-Peter Porsche (v.l.n.r.)
With sons Ferdinand Alexander, Wolfgang and Hans-Peter Porsche (from left to right)

company. The Porsche 356 was shown to the public for the first time in coupe and convertible body styles at the 1949 Geneva Motor Show. A year later, Ferry Porsche had the first cars shipped to America, where they were sold through the dealer network run by Max Hoffman, an automobile importer with a keen eye for business. On the world's largest and most important car market, the Porsche 356 captured the hearts of sports-car enthusiasts – and a large number of Hollywood stars as well. With models such as the 356 Speedster, Ferry Porsche appealed to the taste of customers in the American market, which by 1955 was already taking half of the company's annual output. As well as exports, it was Ferry Porsche's passion for motor sport that helped ensure the brand's success. His view was that instead of trying to convince the public by means of advertising, his cars should demonstrate their merits directly by winning races: "The severe stresses that the cars have to withstand in motor racing soon reveal any weak points and oblige the engineers to look for new and better solutions." This principle resulted in fast, reliable cars that found favor with enthusiasts all over the world. "There was no need for us to encourage

114|115 **DER ERSTE PORSCHE-SPORTWAGEN** | THE FIRST PORSCHE SPORTS CAR

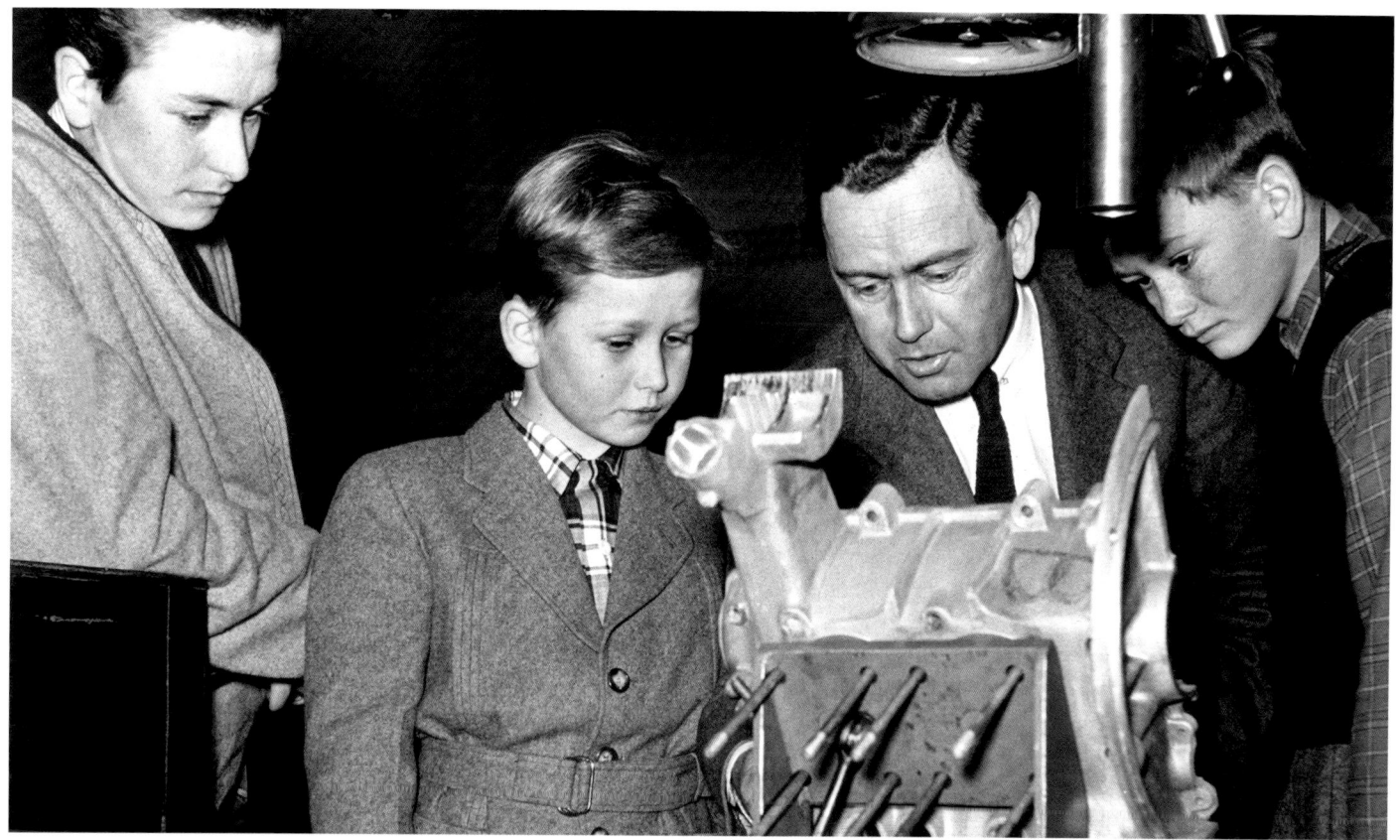

Sportfahrern geschätzt wurden. »Wir brauchten all die Kunden gar nicht zu animieren, sich an Rennen und Rallyes zu beteiligen. Mit einem Fahrzeug dieser Art lag es auf der Hand, sich im Wettkampf mit anderen zu messen.« Der Rennsport bedeutete für Ferry Porsche ständigen technischen Fortschritt, denn die dort gesammelten Erfahrungen flossen direkt in die Perfektionierung der Serienmodelle ein. »Rennen sind praktisch Kriegszustand für die Konstrukteure, und der Druck dieses Notstands beschleunigt die Entwicklung«, fasste er seine Philosophie zusammen. Dieses Konzept ging auf, und so konnte Ferry Porsche bereits 1956 auf mehr als 400 Rennsiege zurückblicken. Der Name Porsche wurde zu einem Synonym für Sportlichkeit, und die Zuffenhausener Sportwagen erhielten schon damals das bis heute charakteristische sportliche Image.

Automobil-Ausstellung Genf 1949 mit Porsche 356/2 »Beutler-Cabriolet« und 356/2 »Gmünd-Coupé«
The 1949 Geneva Motor Show, with the Porsche 356/2 (convertible body by Beutler) and the 356/2 'Gmünd Coupé'

Porsche owners to take part in races or rallies. With a car like this at their disposal, the temptation to try one's luck in competition was impossible to resist!" For Ferry Porsche, motor sport was the key to ongoing technical progress, with the experience that the company gained from it being fed back all the time as improvements to production cars. His philosophy, summed up in brief: "Racing is like a state of war for designers – an emergency situation that exerts pressure on them and speeds up development work." More than 400 race wins by 1956 were proof that Ferry Porsche's approach was correct. The Porsche name became a synonym for sports cars with character, and the products from Zuffenhausen acquired the sporty image that they have retained to this very day.

Ferry Porsche und Max Hoffman im Dezember 1951 auf der Terrasse von Max Hoffmans Wohnung in der Park Avenue in New York
Ferry Porsche and Max Hoffman in December 1951, on the terrace of Hoffman's apartment on Park Avenue, New York

LINKE SEITE: Nach der Rallye Mitternachtssonne 1950 auf dem Killesberg in Stuttgart: Joachim Erbprinz zu Fürstenberg, Ferry Porsche, Constantin Graf von Berckheim, Günther Graf von Hardenberg, Friedrich Prinz zu Fürstenberg

LEFT PAGE: At Killesberg Park in Stuttgart, after the 1950 Midnight Sun Rally: Hereditary Prince Joachim zu Fürstenberg, Ferry Porsche, Constantin Count von Berckheim, Günther Count von Hardenberg and Friedrich Prince zu Fürstenberg

LINKS: Ferry Porsche mit seinem Sohn Wolfgang bei einer Versuchsfahrt (1954)

RECHTS: Der Viernockenwellenmotor Typ 547, auch »Fuhrmann-Motor« genannt, wurde zur Grundlage zahlreicher Porsche-Rennerfolge

LEFT: Ferry Porsche with his son Wolfgang, on a test run (1954)

RIGHT: The Type 547 engine with four overhead camshafts, also known as the 'Fuhrmann engine', was the basis for many Porsche racing successes

Vor dem 24 Stunden-Rennen von Le Mans am 28. Juli 1956
Before the start of the Le Mans 24-hour race on July 28, 1956

Nach der Rückkehr von der 21. Mille Miglia 1954: Herbert Linge, Richard von Frankenberg, Ferry Porsche, Hans Herrmann (von links)
After returning from the 21st Mille Miglia road race, 1954: Herbert Linge, Richard von Frankenberg, Ferry Porsche and Hans Herrmann (from left)

LINKE SEITE: Ferry Porsche (mit karierter Schirmmütze), Mitarbeiter Hans Klauser und Wolfgang Porsche an der Box während des 24-Stunden-Rennens in Le Mans 1956
LEFT PAGE: Ferry Porsche (wearing a checkered peaked cap), departmental employee Hans Klauser and Wolfgang Porsche in the pits at the 1956 Le Mans 24-hour race

Bei den 24 Stunden von Le Mans 1960
At the 1960 Le Mans 24 hour race

Le Mans 1970: Ferry Porsche und Jo Siffert
Le Mans 1970: Ferry Porsche and Jo Siffert

LINKE SEITE: Targa Florio 1969: Ferry Porsche und seine Werksfahrer
LEFT PAGE: *Targa Florio 1969: Ferry Porsche and his race drivers*

Le Mans 1981: Derek Bell (links), Jacky Ickx (mitte) und Ferry Porsche (rechts)
Le Mans 1981: Derek Bell (left), Jacky Ickx (center) and Ferry Porsche (right)

RECHTE SEITE: 1984 am Nürburgring
RIGHT PAGE: *At the Nuerburgring, 1984*

DER ERSTE PORSCHE-SPORTWAGEN | THE FIRST PORSCHE SPORTS CAR

DER UNTERNEHMER FERRY PORSCHE
FERRY PORSCHE – THE BUSINESSMAN

In der Montagehalle Werk II in Stuttgart-Zuffenhausen vor Fahrzeugen der Typenreihe 356 A Convertible D und 356 A Coupé (1958)
At the # 2 assembly plant in Stuttgart-Zuffenhausen, showing Type 356 A Convertible D and 356 A Coupé cars (1958)

DER UNTERNEHMER FERRY PORSCHE
FERRY PORSCHE – THE BUSINESSMAN

Ferry Porsche verstand es stets, die Zeichen der Zeit richtig zu deuten und Marktveränderungen zu erkennen. Ende der 50er Jahre bestanden für den 356, der trotz aller »Fitness«-Programme seine Verwandtschaft mit dem VW-Käfer nicht verleugnen konnte, nur noch geringe Zukunftsaussichten. Statt einer Weiterentwicklung des bewährten Modells entschied sich Ferry Porsche für eine Neukonstruktion, die sich am bewährten Porsche-Konzept mit dem luftgekühlten Boxer-Heckmotor orientieren sollte. Keine leichte Aufgabe, denn der 356 war nach nur eineinhalb Jahrzehnten längst zum Klassiker geworden. Anfang der 60er Jahre waren drei der vier Söhne von Ferry Porsche, der mittlerweile bereits den ersten Enkel in den Armen gehalten hatte, dem väterlichen Vorbild gefolgt und in der Automobilbranche aktiv. Allen voran

Auf der Internationalen Automobilausstellung in Frankfurt 1955 präsentiert Ferry Porsche Bundespräsident Theodor Heuss den Porsche Typ 597 »Jagdwagen«
At the 1955 German Motor Show in Frankfurt, Ferry Porsche shows the Porsche Type 597 'Jagdwagen' ('Hunters' Vehicle') to Germany's President Theodor Heuss

Ferry Porsche always interpreted new trends and identified market changes correctly. By the end of the 1950s the Porsche 356, which despite undergoing all kinds of 'fitness training' could no longer deny its fundamental origins in the VW 'Beetle', Its future market prospects were not very promising, but instead of relying on further development of this admittedly successful model, Ferry Porsche decided that a totally new design was needed, though one that retained the well-proven Porsche concept with an air-cooled, horizontally opposed engine at the rear. A challenging task, since in only fifteen years the 356 had acquired classic status. In the early 1960s, three of Ferry Porsche's four sons had followed their father's example and were working in the car industry. (Ferry Porsche had also had the pleasure of holding

Feier des 25-jährigen Bestehens der Firma Porsche und der Fertigstellung des 10.000. Porsche-Wagens am 16. März 1956
Celebrating the Porsche company's 25th anniversary and production of the 10,000th Porsche car on March 16, 1956

RECHTE SEITE: Vor einem Porsche 356 B Coupé in der Abteilung Wagenauslieferung im Fabrikhof Werk II in Stuttgart-Zuffenhausen (1960)
RIGHT PAGE: *In front of a Porsche 356 B Coupé in the car delivery department in the yard of plant #2 in Stuttgart-Zuffenhausen (1960)*

Ferdinand Alexander, der als Konstrukteur in der Modell-Abteilung des Unternehmens arbeitete. Mit ihm erarbeitete Ferry Porsche die Stilistik des 356-Nachfolgers, der ursprünglich 901 heißen sollte. Ferdinand Alexander Porsche berichtete über die gemeinsame Arbeit: »Als ich damals den 911 konstruiert hatte, stand er von Anfang an hinter mir. Aber nicht, weil ich sein Sohn war, sondern weil er überzeugt war. Er hatte immer ein ausgeprägtes Formgefühl; extreme Farben und Formen mochte er nie.«

Auf der Internationalen Automobil-Ausstellung in Frankfurt präsentierte Porsche 1963 den neuen Sportwagen. Der 911 unterschied sich in vielen Punkten von seinem Vorgänger, nicht nur durch den drehfreudigen Sechszylinder-Motor. Und Ferry Porsche freute sich, dass er »endlich problemlos ein Golfset unterbringt«. Die Entscheidung für den von seinem ältesten Sohn gestalteten 911 war visionär, unterschied er sich doch sowohl in seiner Stilistik wie auch technisch grundlegend von allem bisher Dagewesenen. Auch unternehmerisch war die Einführung des Typ 911 ein großes Wagnis für Ferry Porsche. Im Zuge der Produktionsvorbereitung der neuen Baureihe übernahm

Huschke von Hanstein, Joakim Bonnier und Ferry Porsche beim Formel 1-Rennen auf der Solitude
Huschke von Hanstein, Joakim Bonnier and Ferry Porsche at the Solitude race track

his first grandson in his arms.) Ferdinand Alexander, who worked as a designer in the company's modelmaking department, was an obvious choice to cooperate with his father on styling a successor to the 356. This was the Porsche 911, though it was originally to be called the 901. Ferdinand Alexander Porsche has commented on the joint task as follows: "When I sketched out the 911, he supported me from start to finish. Not just because I happened to be his son, but because he was convinced by what I was doing. He always had a pronounced sense of style, but disapproved of brash colors and extreme forms."

At the 1963 International Motor Show in Frankfurt, Germany, Porsche exhibited its new sports car for the first time. The 911 differed from the existing car in very many ways, not least on account of its free-revving six-cylinder engine. There was more space available too: Ferry Porsche was delighted "to have enough room for my golf clubs at last!" The decision to develop the car styled by his eldest son was visionary in character: both in appearance and technical specification, the 911 departed in almost every respect from what had gone before. Introducing

Ferry Porsche mit seinem ältesten Sohn Ferdinand Alexander am Heck eines Porsche A Carrera Hardtop (1958)
Ferry Porsche with his eldest son Ferdinand Alexander, at the back of a Porsche A Carrera Hardtop (1958)

Seit Oktober 1962 wurden Fahrwerksuntersuchungen auf dem Weissacher Skid-Pad durchgeführt

Since October of 1962 the proving ground in Weissach was to include a skid pad for suspension testing

LINKS: Die nächste Generation: Ferry und Ferdinand Alexander Porsche im Porsche Design-Studio (um 1959)
LEFT: *The next generation: Ferry and Ferdinand Alexander Porsche in the Porsche Design Studio (about 1959)*

Im Oktober 1961 auf dem Titel des Magazin »Der Spiegel«
Front cover of the magazine 'Der Spiegel' in October 1961

the 911 was naturally a considerable risk for Porsche. As part of the preparations for the new model line, Porsche took over the body supplier Karosseriewerk Reutter & Co. GmbH in 1964. This in itself called for a major effort n the part of the sports-car manufacturer: almost a thousand Reutter employees were taken over by Porsche KG with their previous job tenure taken into account. One of Ferry Porsche greatest achievements is surely to have identified the future potential of such an innovative concept as the 911. The success of the new model line was based on steady evolution, which made the 911 the perfect sports car. Ferry Porsche: "When I look back to development of the 911, it certainly wasn't a universally accepted concept. But the exceptionally long career of this model makes me proud that my opinion of the car proved correct in the long run."

Ferry Porsche often gave unusual, even risky, ideas a chance to prove their worth. In 1969 the company cooperated with Volkswagen on the VW-Porsche 914, a low-priced sports car intended to capture market share below the Porsche 911. Although its styling and image were controversial, more than 120,000 were

Ferry Porsche 1963 (auf dem Kotflügel des Porsche 904 sitzend) im Kreise von Porsche-Händlern aus USA anlässlich einer Besichtigung in Stuttgart-Zuffenhausen

Ferry Porsche 1963 (seated on the fender of the Porsche 904) with Porsche dealers from the USA during their visit to Stuttgart-Zuffenhausen

Ferry Porsche und der
Porsche 911 (1968)
*Ferry Porsche and the
Porsche 911 (1968)*

Familienrat am 911 S 2,0 Targa: (v.l.): Ferdinand Piëch, Ferry Porsche, Hans-Peter Porsche, Ferdinand Alexander Porsche (1968)
Family council with 911 S 2.0 Targa (from left): Ferdinand Piëch, Ferry Porsche, Hans-Peter Porsche and Ferdinand Alexander Porsche (1968)

1977: Ferry Porsche während der Feier anlässlich des 250.000. Porsche-Sportwagens, einem 911 2,7 Coupé
1977: Ferry Porsche at the celebration of the 250,000th Porsche sportscar, a 911 2,7 coupé

Ferry Porsche mit einem
Typ 356 C (links) und einen
911 Carrera 4 (1989)
*Ferry Porsche with a
Type 356 C (left) and a
911 Carrera 4, 1989*

Porsche Tennis Grand Prix
1986: Die Siegerin Martina
Navratilova erhielt von
Ferry Porsche ein 911
Carrera 3,2 Cabriolet
*Porsche Tennis Grand
Prix, 1986: winner Martina
Navratilova is presented
with a 911 Carrera
3.2 Convertible by
Ferry Porsche*

Ferry Porsche und
das Sondermodell
»30 Jahre 911« (1993)
*Ferry Porsche and the
special '30th Anniversary'
911 model (1993)*

»30 Jahre 911«:
Ferry Porsche mit seinem
ältesten Sohn Ferdinand
Alexander (1993)
*30th anniversary of the
911: Ferry Porsche with
his eldest son Ferdinand
Alexander (1993)*

Produktionsjubiläum: Am 15. Juli 1996 lief der 1.000.000. Porsche-Sportwagen vom Band. Das Jubiläumsauto erhielt die baden-württembergische Autobahnpolizei
Production jubilee: the millionth Porsche sports car was completed on July 15, 1996, and presented to the highway police in the German State of Baden-Wuerttemberg

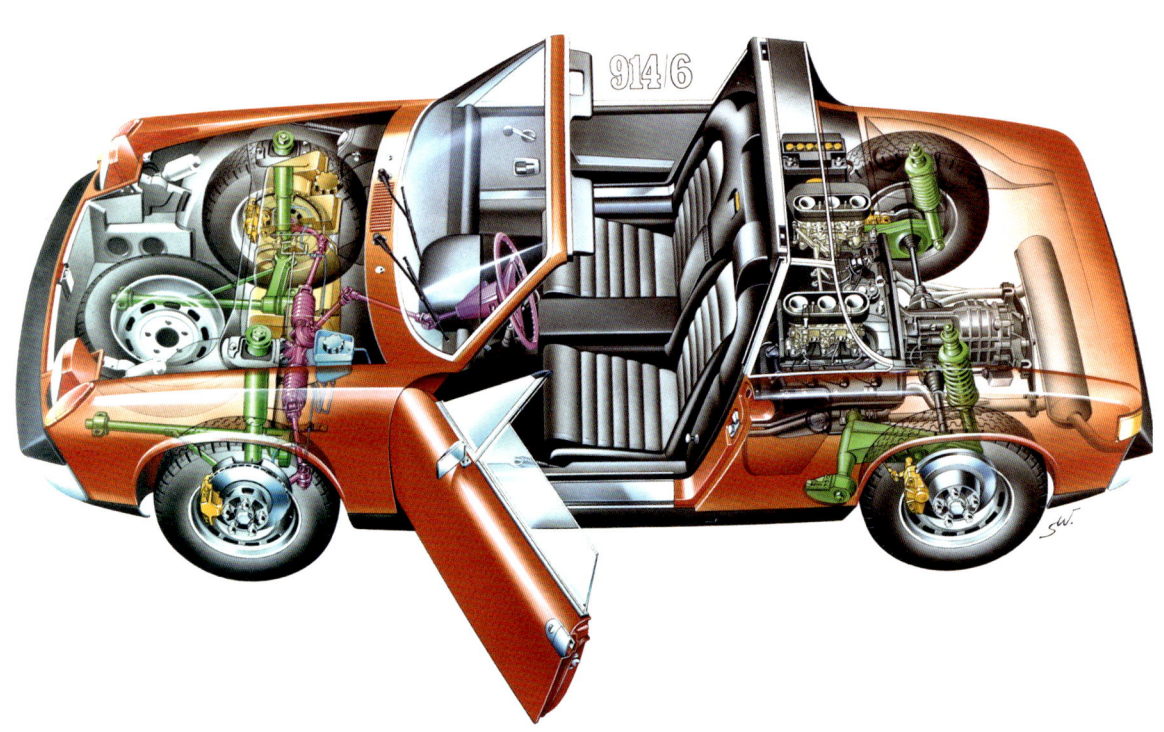

Porsche 1964 das Zuliefererunternehmen Karosseriewerk Reutter & Co. GmbH. Dies bedeutete einen großen Kraftakt für den Sportwagenhersteller, denn die knapp 1.000 Mitarbeiter der Firma Reutter wurden unter Anrechung ihrer Betriebszugehörigkeit komplett in die Porsche KG übernommen. Dass Ferry Porsche das Zukunfts- und Erfolgspotenzial des durch und durch innovativen 911-Konzeptes erkannt hat, ist eines seiner großen Verdienste. Das Erfolgsrezept der Baureihe war die Modellpolitik der beständigen Evolution, die den 911 zum perfekten Sportwagen reifen ließ: »Blicke ich auf den 911 zurück, so stellt dieser Typ zweifellos ein umstrittenes Konzept dar. Die lange, geradezu ungewöhnliche Lebensdauer dieses Modells macht mich doch stolz darauf, am Ende mit meiner Meinung vom 911 recht behalten zu haben.«

Auch ungewöhnlichen und riskanten Ideen gab Ferry Porsche immer wieder eine Chance. Mit dem VW-Porsche 914 ging das Unternehmen 1969 eine Kooperation mit Volkswagen ein, um mit einem preisgünstigen Sportwagentyp neue Marktanteile unterhalb des Porsche 911 zu gewinnen. Obwohl auf den Gebieten Design und Image nicht unumstritten, wurde auch der

LINKE SEITE: Phantomzeichnung eines VW-Porsche 914/6 (1970)
LEFT PAGE: *Ghosted drawing of the VW-Porsche 914/6 (1970)*

Internationale Automobilausstellung Frankfurt 1969: Ferry Porsche im Gespräch mit Bundeskanzler Kurt Georg Kiesinger und Volkswagenwerk-Chef Kurt Lotz an einem VW-Porsche 914
The German Motor Show in Frankfurt, 1969: Ferry Porsche talking to Germany's Chancellor Kurt Georg Kiesinger and Volkswagen's CEO Kurt Lotz at a VW-Porsche 914

Ferry Porsche mit seinem Porsche 914/8 auf dem Schüttgut, dem Familiensitz der Familien Porsche und Piëch. Das Fahrzeug wurde ihm zum 60. Geburtstag als Geschenk übergeben
Ferry Porsche with his 914/8 at the 'Schüttgut', the Porsche and Piëch families' property. The car was his 60th birthday present

RECHTE SEITE: Diesen 928 S erhielt Ferry Porsche 1979 von seiner Belegschaft zum 70. Geburtstag
RIGHT PAGE: *Ferry Porsche was presented with this 928 S for his 70th birthday in 1979*

154|155 DER UNTERNEHMER FERRY PORSCHE | FERRY PORSCHE – THE BUSINESSMAN

LINKE SEITE: Ein viersitziger Porsche 928 S war das Geschenk der Mitarbeiter anlässlich des 75. Geburtstags 1984
LEFT PAGE: *The workforce gave Ferry Porsche a four-seat Porsche 928 S on the occasion of his 75th birthday in 1984*

RECHTE SEITE: Die Studie Panamericana erhielt Ferry Porsche 1989 zum 80. Geburtstag. Hier mit seinen Söhnen Hans-Peter, Gerhard, Ferdinand Alexander und Wolfgang auf Schloss Prielau bei Zell am See
RIGHT PAGE: *Ferry Porsche was given this 'Panamericana' study in 1989, to celebrate his 80th birthday. He is seen here with his sons Hans-Peter, Gerhard, Ferdinand Alexander and Wolfgang at Prielau Castle near Zell am See, Austria*

Mit Louise Piëch am 60. Geburtstag von Ferry Porsche auf dem Schüttgut in Zell am See (1969)
Together with Louise Piëch at Ferry Porsche's 60th birthday at the 'Schüttgut' in Zell am See, 1969

914 zu einem Erfolg und mit fast 120.000 Exemplaren der erfolgreichste Sportwagen der frühen 70er Jahre. Auch Ferry Porsche fuhr einen 914, den er als Geschenk seiner Belegschaft zum 60. Geburtstag erhielt. In dem äußerlich fast unveränderten Sportwagen arbeitete allerdings der Dreiliter-Achtzylinder des Rennsportwagens Typ 908 in einer für den Alltag gezähmten Version. 260 PS standen im Einzelgutachten des Wagens mit der offiziellen Zulassungsnummer S-R 3000. Es war kein Geschenk für die Garage: Über 10.000 Kilometer spulte Ferry Porsche mit dem 914/8 ab. Von da ab wurde das individuelle Geburtstagsauto bei runden Geburtstagen des Porsche-Chefs zur Tradition. Auf die Frage, ob er sich denn selber auch einen Porsche kaufen würde, antwortete Ferry Porsche einem Journalisten: »Nein, ich warte einfach bis ich Geburtstag habe.«

Nicht nur auf technischer Ebene, sondern auch auf sozialem Gebiet wollte Ferry Porsche sein Unternehmen ganz nach vorne bringen. Unternehmer sein und Mensch bleiben war für ihn kein Gegensatz, sondern eine logische Ergänzung, vielleicht sogar Basis allen Erfolgs. Schon 1956 führte er die betriebliche

1969: Ferry Porsche und der Typ 917. Im Hintergrund (v.l.): Ferdinand Piëch, Hans-Peter Porsche, Ferdinand Alexander Porsche
1969: Ferry Porsche and the type 917. In the backround (from left): Ferdinand Piëch, Hans-Peter Porsche, Ferdinand Alexander Porsche

sold, making the 914 the most successful sports car of the early 1970s. Ferry Porsche drove a 914 that the workforce had presented to him on his 60th birthday – but a rather special one. Externally almost unchanged in appearance, it was powered by the three-liter, eight-cylinder engine from the Type 908 sports racing car, detuned to 260 brake horsepower to make it suitable for road use. This one-off car was

Wagenabholung des Stuttgarter Oberbürgermeisters Arnulf Klett, 1969. Rechts: Arnulf Klett; 3.v.l. Ferry Porsche, 4.v.l. Huschke v. Hanstein; Porsche Typ 911 E 2,2 Coupé
Arnulf Klett, Mayor of Stuttgart, collecting a car in 1969. At right: Arnulf Klett; 3rd from left: Ferry Porsche; 4th from left: Huschke v. Hanstein: the car is a Porsche 911 E 2.2 Coupe

Unternehmer Ferry Porsche am 10. Mai 1974 anlässlich der Feier zum Jubiläum »25 Jahre Porsche-Produktion«
Ferry Porsche the businessman, during the '25 Years of Porsche Production' jubilee on May 10, 1974

granted a special road-use permit, and bore a license plate from the city of Stuttgart with the number S–R 3000. It spent very little time in the garage: Ferry Porsche covered more than 10,000 kilometers in the 914/8. From then on, it became a tradition for the workforce to present Porsche's chief executive officer on significant birthdays with a car that was unique in some way. Asked once by a journalist whether he would buy one of his own cars, Ferry Porsche replied with a hint of irony: "No, I just wait until my birthday comes around!"

It was Ferry Porsche's aim for his company to be right up among the leaders not only technologically but also in the field of social welfare. For him, business talent and humanity were entirely compatible, in fact the one was a logical complement to the other and possibly even the only joint basis for success. He introduced a company old-age pension scheme in 1956 and established a Porsche Foundation that assisted any employees who found themselves in a precarious economic situation through no fault of their own. As long ago as 1960 the company transferred all its workforce from an hourly-wage to a monthly-salary

Altersversorgung ein, und eine Porsche-Stiftung half fortan allen Mitarbeitern, die unverschuldet in eine wirtschaftliche Notlage geraten waren. Schon 1960 übernahm das Unternehmen alle Arbeiter vom Stundenlohn in den Monatslohn und stellte sie damit den Angestellten gleich. Auch Leistungen wie Weihnachts- oder Urlaubsgeld sowie Lohnfortzahlung im Krankheitsfall wurden bei Porsche weit früher – und ohne gesetzliche oder tarifliche Verpflichtung – eingeführt, als es in der Branche üblich war. Ferry Porsche war damit auch ein Pionier auf dem Gebiet der unternehmerischen Sozialleistungen. Als Unternehmer verstand er es zudem, seinen Mitarbeitern viel Freiraum zu geben und sie auf diese Weise zu Höchstleistungen zu motivieren. Über seine Führungsphilosophie sagte er: »Ich habe von meinem Vorbild, meinem Vater, der sehr streng war und seine Mitarbeiter manchmal hart anfasste, gelernt, dass man mit ärgerlich sein nicht so sehr viel erreicht, damit einen Mitarbeiter eher zur Verzweiflung treibt. Ich habe eigentlich immer versucht, solange ich hier allein maßgeblich war, alle meine Herren an langer Leine laufen zu lassen, denn man soll doch dem einzelnen, wenn er kreativ ist, nicht unbedingt etwas aufzwingen wollen.«

Ferry Porsche und Ernst Fuhrmann (rechts), ab August 1972 Sprecher des Vorstandes, von 1976 bis 1980 Vorsitzender des Vorstandes der Porsche AG
Ferry Porsche and Ernst Fuhrmann (right); Fuhrmann became Executive Board Spokesman in August 1972 and was Chairman of the Porsche AG Executive Board from 1976 to 1980

basis, so that the distinction between wage-earners and salaried staff was abolished. Porsche also introduced bonuses such as Christmas or vacation pay well before the industry in general, and without any legal compulsion or negotiations with the trade unions. Ferry Porsche's company was a definite pioneer in the social welfare area. As a businessman, he understood that his employees had to be given plenty of scope for personal development if they were to be motivated for high performance. Commenting on his management philosophy, he once said: "From my father, who was an example to me but nevertheless very strict and inclined to dish out hard treatment to his employees, I learned that getting annoyed with people doesn't achieve very much, but tends to make them feel that they're in a hopeless situation. For as long as I was in charge at the company, I tried to give my staff free rein, and if people were creative not to compel them always to do something different."

Ferry Porsche had the interests of his workforce at heart, but those of his customers too. He took their wishes and suggestions very seriously, and when he attended Porsche Club meetings tried to make direct

Porsche-Parade am 21. September 1984 anlässlich des 75. Geburtstages von Ferry Porsche
Porsche Parade on the occasion of Ferry Porsche's 75th birthday, September 21st, 1984

Den Aufsichtsratsvorsitz übergab Ferry Porsche 1990 an seinen ältesten Sohn Ferdinand Alexander
In 1990 Ferry Porsche handed over Chairmanship of the Supervisory Board to his eldest son Ferdinand Alexander

Der Ehrenvorsitzende des Aufsichtsrats Ferry Porsche und der Vorstandsvorsitzende Dr. Wendelin Wiedeking auf der Porsche-Hauptversammlung 1997
Ferry Porsche, Honorary Chairman of the Supervisory Board, and CEO Dr. Wendelin Wiedeking at the Porsche annual general meeting in 1997

Neue Wege: Ferry Porsche und der 928 (1978)
A new departure: Ferry Porsche and the 928 (1978)

Bestseller: Ferry Porsche und der 100.000. Porsche 924 (1981)
Best-seller: Ferry Porsche and the 100,000th Porsche 924 (1981)

Mit dem Porsche 968 Coupé in der Zuffen-hausener Produktion (1991)
The Porsche 968 Coupé at the Zuffenhausen production plant (1991)

Neben den Mitabeitern waren es vor allem die Kunden, die Ferry Porsche am Herzen lagen. Ihre Wünsche und Anregungen nahm er stets sehr ernst und suchte auf Treffen der zahleichen Porsche-Clubs den direkten Kontakt zu den Käufern seiner Automobile. Nur zu gerne prämierte oder gar signierte er die von ihren Besitzern liebevoll gepflegten Fahrzeuge. Dies war seine Art dafür Respekt zu zollen, dass viele seiner Kunden viel Zeit und Geld in Ihre Porsche-Sportwagen investierten.

WEICHENSTELLUNG FÜR DIE ZUKUNFT

Anfang der siebziger Jahre bestimmte Ferry Porsche noch einmal die langfristige Richtung für das schon über zwei Jahrzehnte von ihm geleitete Unternehmen. Nach ausgiebigen Diskussionen um seine Nachfolge in der Porsche-Geschäftsführung folgte 1971 der Beschluss der Familien Porsche und Piëch, zukünftig keine operativen Führungspositionen mehr mit Familienangehörigen zu besetzen. Zum Jahresbeginn 1972 vereinbarten die Gesellschafter der »Dr. Ing. h.c. F. Porsche KG«, die Kommanditgesellschaft zum 1. August 1972 in eine Aktiengesellschaft umzuwandeln.

contact with the people who bought his cars. He loved to present prizes to cars that had been treated with immense care by their owners, and even used to autograph them. It was his way of showing his respect for the customers who invested so much time and money in their Porsche sports cars.

SETTING THE SIGNALS FOR THE FUTURE

In the early nineteen-seventies Ferry Porsche once again made the decisions that governed the long-term policy of the company he had managed for more than twenty years. After extensive discussion of his successor at the head of the Porsche Executive Board, the Porsche and Piëch families decided in 1971 that in the future operative management positions should no longer be filled by members of the families. At the beginning of 1972 the partners in Dr. Ing. h.c. F. Porsche KG voted to convert the existing private limited partnership into a joint stock corporation with August 1, 1972 as the deadline. The equity, amounting to 50 million Deutschmarks, was held in equal parts by members of the Porsche and Piëch families.

Autogrammstunde:
Ferry Porsche beim
Signieren eines 944 Turbo
(1986)
*Signature session: Ferry
Porsche autographs a 944
Turbo (1986)*

Ferry Porsche 1984 in
seinem Arbeitszimmer vor
dem Bild seines Vaters
Ferdinand Porsche.
*Ferry Porsche in his
office in 1984, in front
of a picture of Ferdinand
Porsche, his father*

Das Grundkapital in Höhe von 50 Millionen DM wurde zu gleichen Teilen von den Familienstämmen Porsche und Piëch gehalten.

Auch Ferry Porsche unterwarf sich dem einstimmigen Familienbeschluss und zog sich aus der aktiven Geschäftsführung zurück, um fortan als Vorsitzender des Aufsichtsrates die Geschicke des Unternehmens zu begleiten. Dieses Amt übte er bis ins Jahr 1990 aus, als sein Sohn Ferdinand Alexander den Vorsitz von ihm übernahm. Als Ehrenvorsitzender des Aufsichtsrats sollte Ferry Porsche dann die Entwicklung des Unternehmens bis an sein Lebensende aktiv verfolgen.

Sein unternehmerischer und technischer Weitblick beeinflussten jedoch das Unternehmen Porsche immer wieder noch nachhaltig. Gerne ließ er sich von den innovativen Ideen seiner Ingenieure begeistern und überzeugen. Mit Modellen wie dem 924, 944 und 928 verließ Porsche gewohnte Fahrwasser – nicht immer zur Freude der »gusseisernen« Kunden.

Zum Unternehmenserfolg trugen aber auch diese Modelle ganz erheblich bei, denn immerhin war in den 80er Jahren jeder zweite produzierte Porsche ein solcher Frontmotor-Sportwagen. Viel beschäftigte

Ferry Porsche accepted this unanimous vote and withdrew from active company management, though as Chairman of the Supervisory Board he naturally continued to watch over the affairs of the company. He held this post until 1990, after which his place was taken by his son Ferdinand Alexander. As Honorary Chairman of the Supervisory Board, Ferry Porsche continued to play an active part in the company's progress until the end of his life.

His far-sighted business and technical policies continued on many occasions to exert a powerful influence on the Porsche company. He studied its engineers' innovative ideas and backed them enthusiastically when he was convinced of their merits. With its 924, 944 and 928 models, Porsche explored new model concepts – not always with the approval of 'dyed in the wool' customers. None the less, these models made a worthwhile contribution to the company's success: in the 1980s, half of all the Porsche sports cars that left the production lines were of front-engine design. Ferry Porsche also concerned himself in depth with the question of the automobile's future. His assessment, which dates

Ferry Porsche im Porsche
911 Carrera 3,2 Speedster
(1989)
*Ferry Porsche in a Porsche
911 Carrera 3.2 Speedster
(1989)*

RECHTE SEITE: Mit dem Boxster gelang Porsche der wirtschaftliche Turnaround (1997)
RIGHT PAGE: *With the Boxster came the turnaround in Porsche's fortunes (1997)*

Ferry Porsche an seinem
85. Geburtstag (1994)
*Ferry Porsche on his 85th
birthday (1994)*

sich Ferry Porsche aber auch mit der Frage nach der Zukunft des Automobils. Seine Einschätzung aus dem Jahr 1979 erscheint heute aktueller denn je: »In der Zukunft kommt es besonders auf den Verbrauch an. Der Verbrauch ist seit jeher abhängig von Gewicht und Luftwiderstand. In beiden Punkten ist der Sportwagen im Vorteil.« Wichtig sei, dass »wir an unseren Wagen Dinge tun, die für den Verbrauch ausschlaggebend sind. Da nützen uns Entwicklungen, die wir für den Sport vorangetrieben haben – zum Beispiel der Turbolader. Wir können den Turbolader nicht nur nutzen, um die Leistung zu steigern. Wir können mit ihm auch den Wirkungsgrad des Motors verbessern und kommen dann auf extrem niedrige Verbrauchswerte.«

Auch in seinen Vorstellungen über zukünftige Porsche-Modellprogramme war Ferry Porsche häufig seiner Zeit weit voraus. Er ließ seinen Ingenieuren große Freiheiten bei der Konzeption neuer Ideen, über die er bei seinen häufigen Rundgängen durch die Entwicklungsabteilung gerne diskutierte. Von neuen technischen Ideen ließ er sich gerne überzeugen, denn »Porsche kann und darf alles bauen. Das Produkt muss nur besser als alles Vergleichbare sein.«

from 1979, now seems more topical than ever: "The crucial factor in the future will be fuel consumption. This in turn is directly related to weight and resistance to the airflow, both areas in which the sports car has an advantage." Ferry Porsche felt that "we do things to our cars that decisively affect their fuel consumption. We make use of developments that we originally undertook for motor sport, for example the turbocharger. We can use this device to boost power output, but also to increase the engine's efficiency and achieve very low fuel consumption."

In his vision of the Porsche model programs of the future, Ferry Porsche was often well ahead of his time. He granted his engineers plenty of freedom to come up with new ideas, and liked to discuss these when one of his frequent walkabouts took him to the development department. He willingly accepted new technical ideas, and used to say "Porsche can and must build everything that makes the product better than anything it could be compared with!"
In the early 1970s he asked his designers to sketch out an all-wheel-drive vehicle with a six-cylinder engine, which as he put it "would give the company

Auf kurzesten Nenner gebracht ist dieses Modell, das den 2-Liter-Carrera ablöst, in Leistung, Ausstattung und Fahrkomfort ein 2+2-sitziges Coupé europäischer Spitzenklasse.

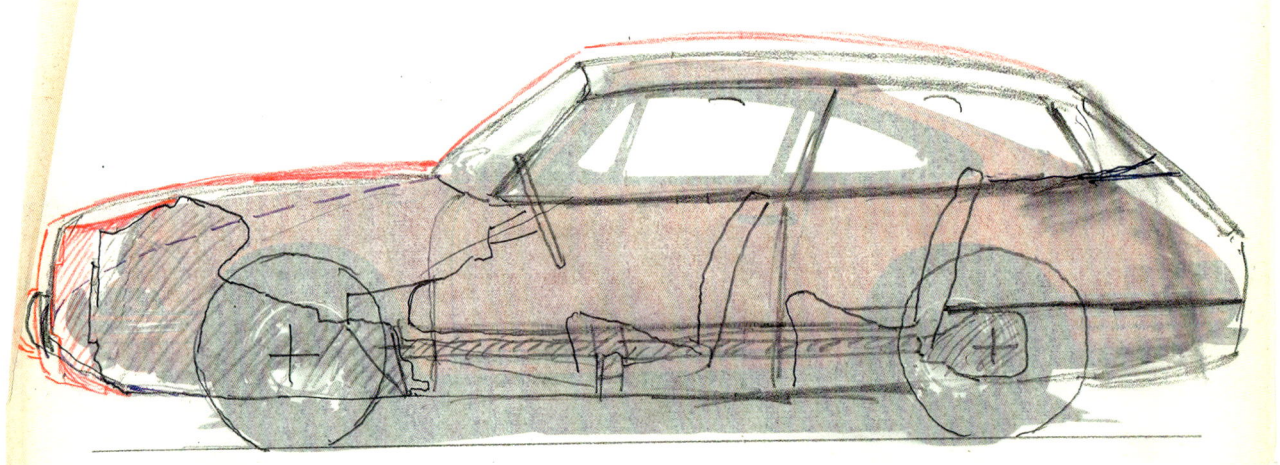

Bereits Anfang der 70er Jahre ließ er seine Konstrukteure ein allradangetriebenes Fahrzeug mit Sechszylindermotor entwerfen, um, so Ferry Porsche, einmal »als zweites Standbein eine Art Range Rover zu bauen.« Vom Erfolg dieser Idee überzeugt, sagte er 1989 in einem Interview: »Wenn wir ein Geländefahrzeug nach unseren Qualitätsvorstellungen bauten, und vorne steht Porsche drauf, würde es auch verkauft.« Verwirklicht wurde diese Version im Jahr 2002, als Porsche mit dem sportlichen und geländegängigen Mehrzweckfahrzeug Cayenne in das Marktsegment der Sports Utility Vehicles (SUV) einzog. Mit seiner Einschätzung über den Erfolg eines geländegängigen Porsche hat Ferry Porsche Recht behalten, denn seit seiner Markteinführung sind inzwischen mehr als eine viertel Million Einheiten des Porsche Cayenne abgesetzt worden.

Eine weitere Vision Ferry Porsches war der Wunsch nach einem viertürigen Porsche. Seit den Tagen des Typ 356 hat er sich mit diesem Thema beschäftigt und im Entwicklungsbereich mehrfach Studien beauftragt. Aus technischen und modellpolitischen Gründen konnte er die Verwirklichung nicht mehr erleben. Mit dem viertürigen Gran Turismo Panamera, der passend

Eigenhändige Zeichnung eines viersitzigen Porsche von Ferry Porsche
Ferry Porsche's drawing of a four-seat Porsche

a kind of Range Rover and a second leg to stand on." He was convinced that this idea would be a success, and even in 1989 told an interviewer: "If we build an off-road model according to our standards of quality, and it has a Porsche badge on the front, people will buy it!" This vision came true in 2002, when Porsche entered the Sports Utility Vehicle (SUV) market segment with the Cayenne. Ferry Porsche's estimate of the potential represented by an off-road Porsche model proved to be absolutely correct; since it was launched, more than a quarter of a million Porsche Cayennes have been sold.

Ferry Porsche had another vision, too: he wanted to build a four-door Porsche. This topic had interested him ever since the early days of the 356, and he asked the development staff more than once to draw up a design study. For technical reasons and model-policy reason, such a car never went into production. Now the Panamera, a true 'gran turismo' model, has reached the market, in time to celebrate the centenary of Ferry Porsche's birth. It is a fitting tribute to the man who established Porsche as a sports-car manufacturer and ran the company for so long.

Ferry Porsche und der 911 GT1 (1997). Kunden konnten eine Kleinserie von 20 Porsche 911 GT1 mit Straßenzulassung ordern
Ferry Porsche and the 911 GT1, 1997. For customers a batch of twenty Porsche 911 GT1 licensed for road use was built

zum 100. Geburtstag von Ferry Porsche auf den Markt kommt, ehrt der Sportwagenhersteller Porsche seinen Gründer und langjährigen Firmenchef.

Im Alter musste Ferry Porsche erleben, wie sein Unternehmen in der zweiten Hälfte der 80er Jahre in eine existenzbedrohende Krise geriet. Doch auch als Porsche als Übernahmekandidat gehandelt wurde, betonte er stets den unbedingten Willen zur Selbstständigkeit. Avancen verschiedener Volumenhersteller, die Porsche gerne als Teil ihres Markenportfolio gesehen hätten, lehnte er entschieden ab: »Ich habe dem Unternehmen nicht meinen Familiennamen gegeben, weil ich es einmal verkaufen will.« Mit dem wirtschaftlichen Turnaround war es ihm vergönnt, mitzuerleben, wie sein Lebenswerk wieder auf die Erfolgsspur zurückfand. 1996 erlebte er mit der Einführung des Porsche Boxster die Fortführung seiner Vision eines Mittelmotorroadsters. Über die Zukunft seiner Sportwagenidee war ihm nicht bange. »Das letzte Auto, das gebaut werden wird, wird ein Sportwagen sein.« Mit seinem Tod am 27. März 1998 endete auch eine andere Ära – im selben Jahr lief der letzte luftgekühlte 911 vom Band.

»Am Anfang schaute ich mich um, konnte aber den Wagen, von dem ich träumte nicht finden. Also beschloss ich, ihn mir selbst zu bauen.«

Dr. Ing. h.c. Ferdinand Anton Ernst »Ferry« Porsche

"I looked around first, but I couldn't find my dream car. So I decided to build it myself!"

Dr. Ing. h.c. Ferdinand Anton Ernst 'Ferry' Porsche

In his old age Ferry Porsche was forced to see his company slipping down into a crisis that threatened its existence. This was in the late 1980s, and before long the company was being pushed by the media as a likely takeover candidate. Ferry Porsche never ceased to emphasize its desire to remain independent. Various high-volume carmakers who would gladly have added Porsche to their product portfolio put out feelers, but Ferry Porsche rejected them all with great determination. "I didn't give the company my name just for it to be sold off later!" Before long the company had turned the corner financially and was on the road to commercial success again. When the Porsche Boxster was introduced in 1996, Ferry Porsche had the pleasure of seeing a version of his mid-engine roadster concept reach the market. He never doubted for a moment that his sports-car concept had a future. "The last car anyone builds will be a sports car!" When Ferry Porsche died on March 27, 1998, another era was also coming to an end – in the same year, the last Porsche 911 with an air-cooled engine left the assembly line.

LEBENSLAUF
PERSONAL CHRONICLE

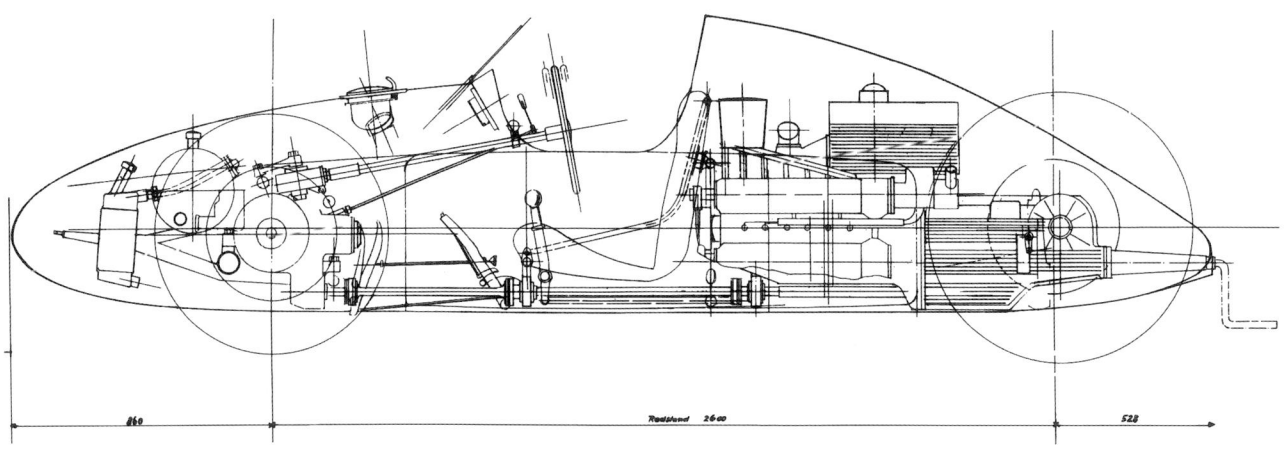

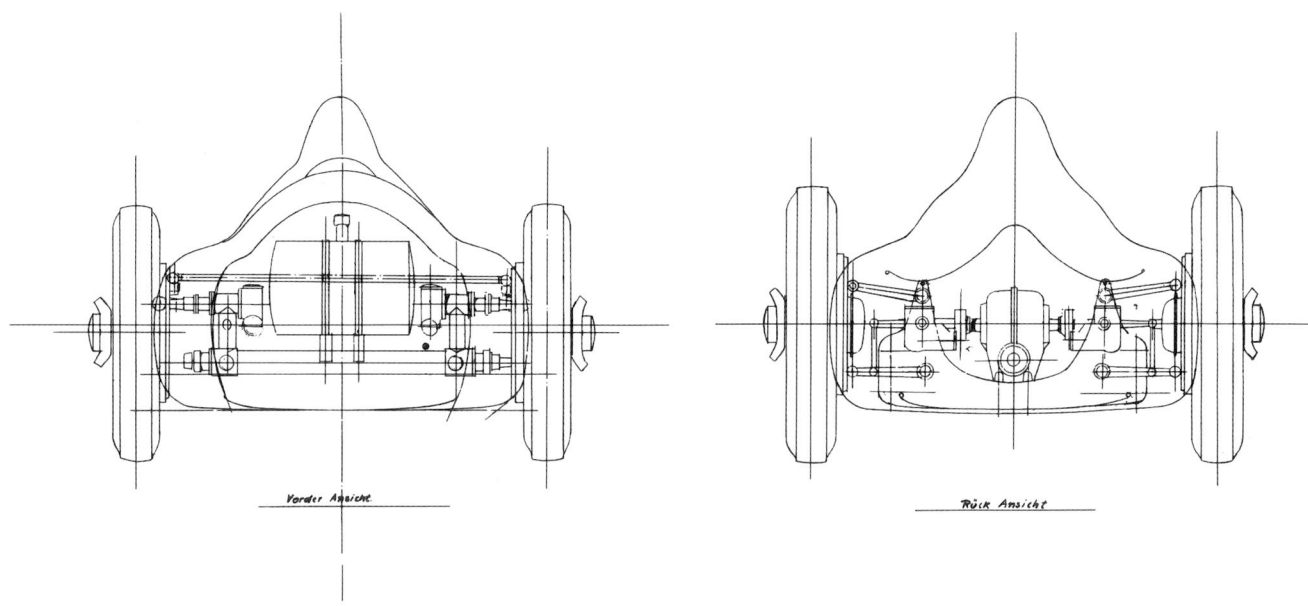

Zeichnung Typ 360 »Cisitalia« vom 10. August 1947
Drawing of the Type 360 'Cisitalia', August 10, 1947

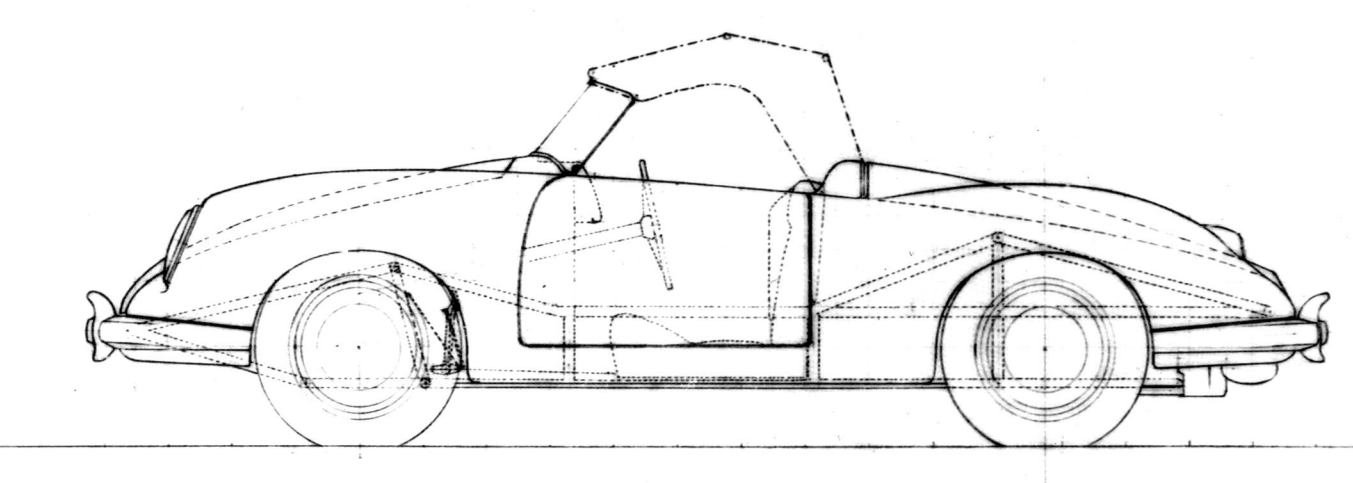

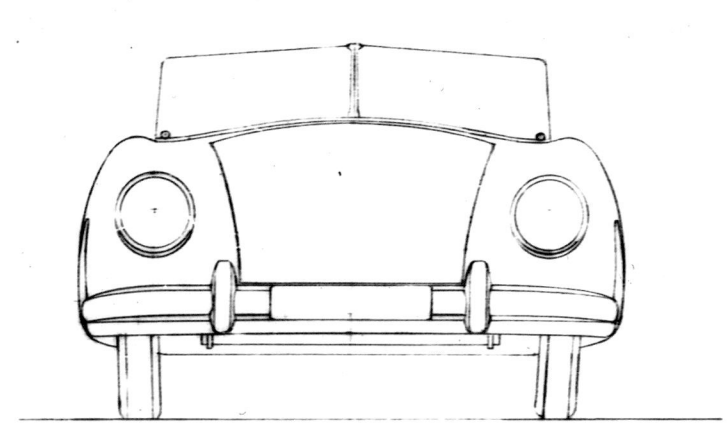
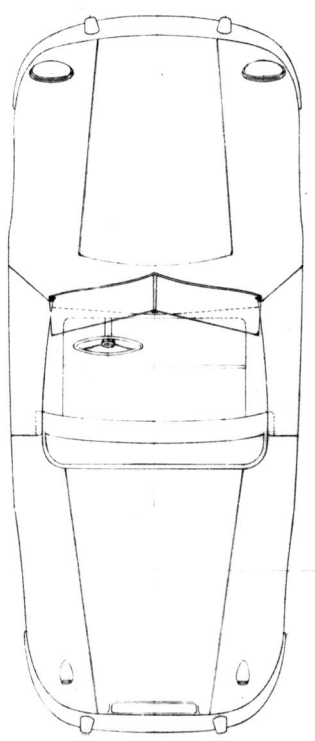

Zeichnung Typ 356 »Nr.1« vom 6. Januar 1948
Drawing of the Type 356 'No. 1', January 6, 1948

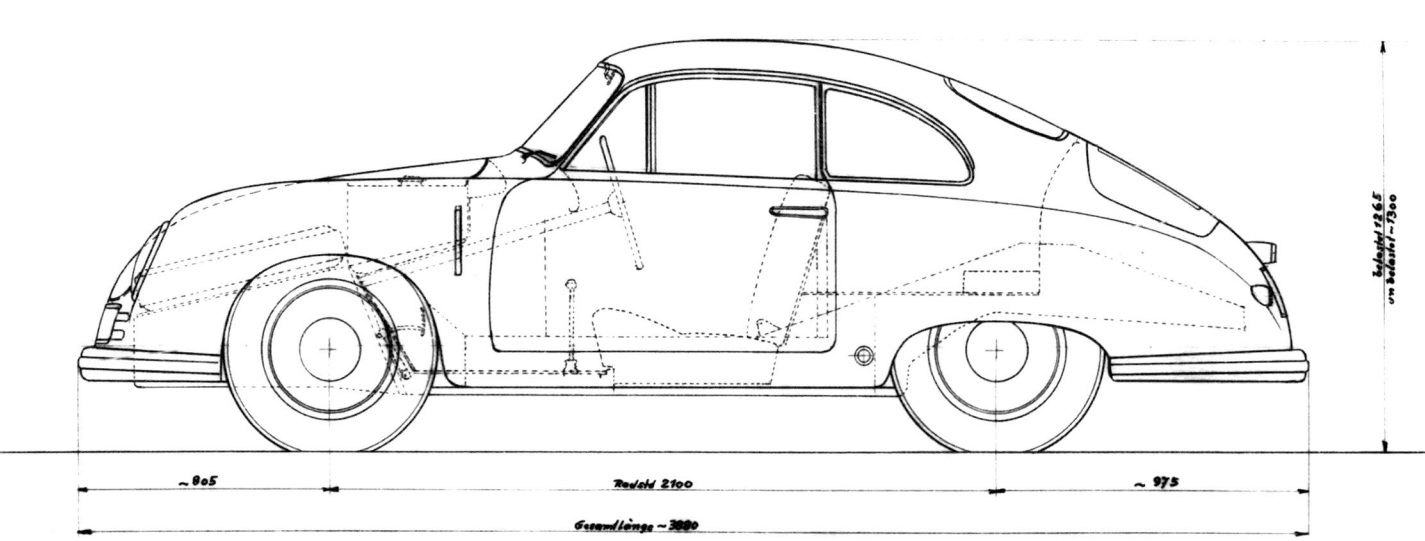

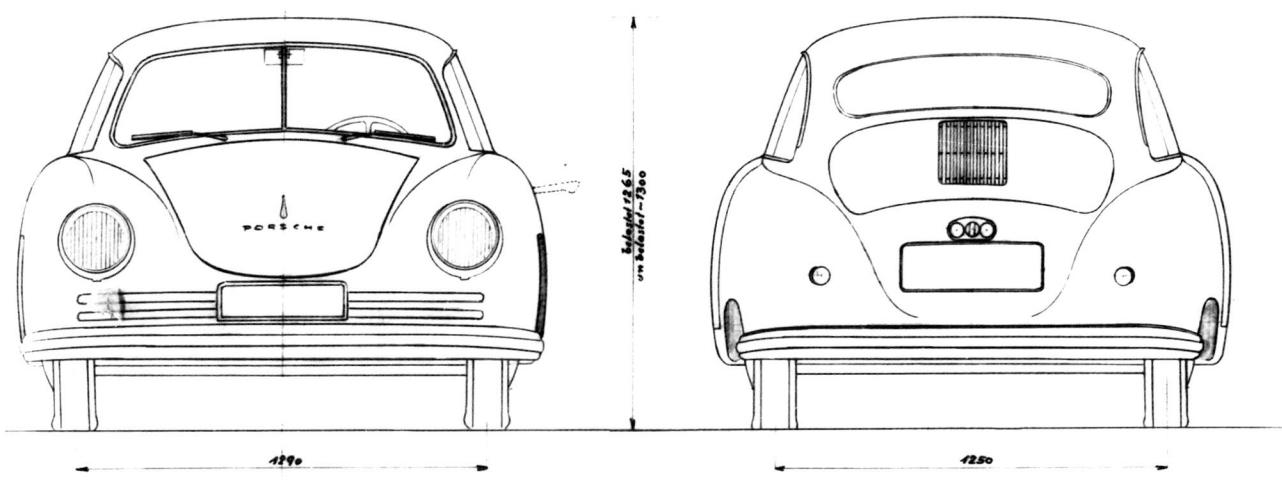

Zeichnung Typ 356/2 »Gmünd-Coupé« vom 25. August 1949
Drawing of the Type 356/2 'Gmünd Coupé', August 25, 1949

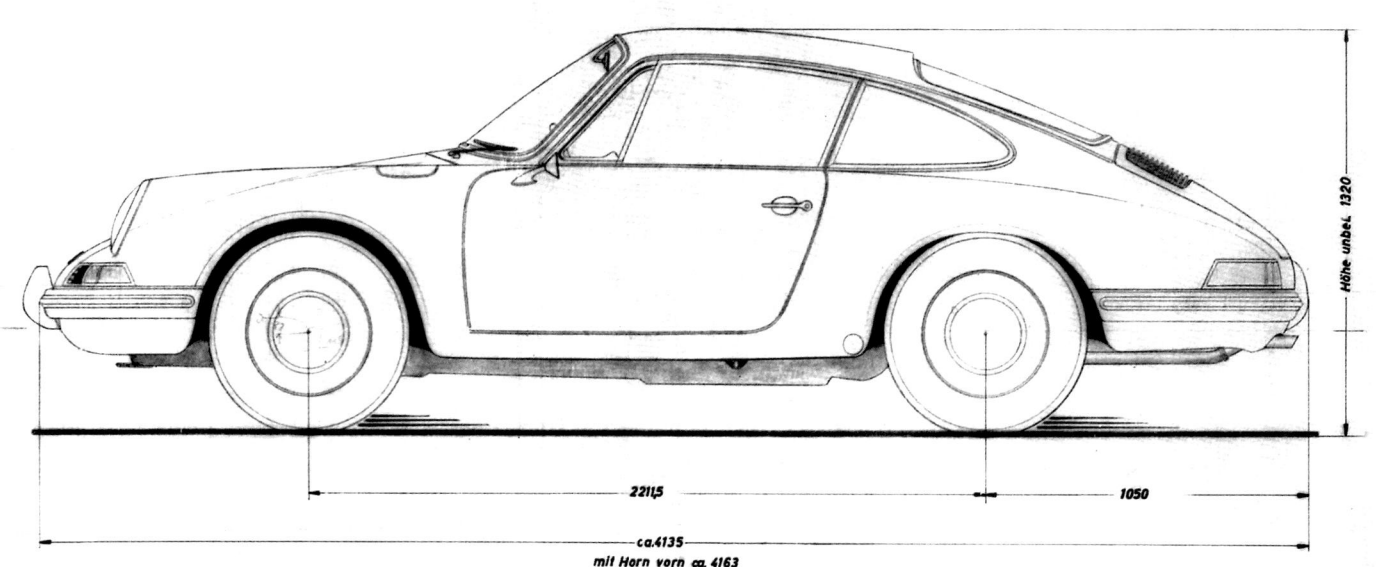

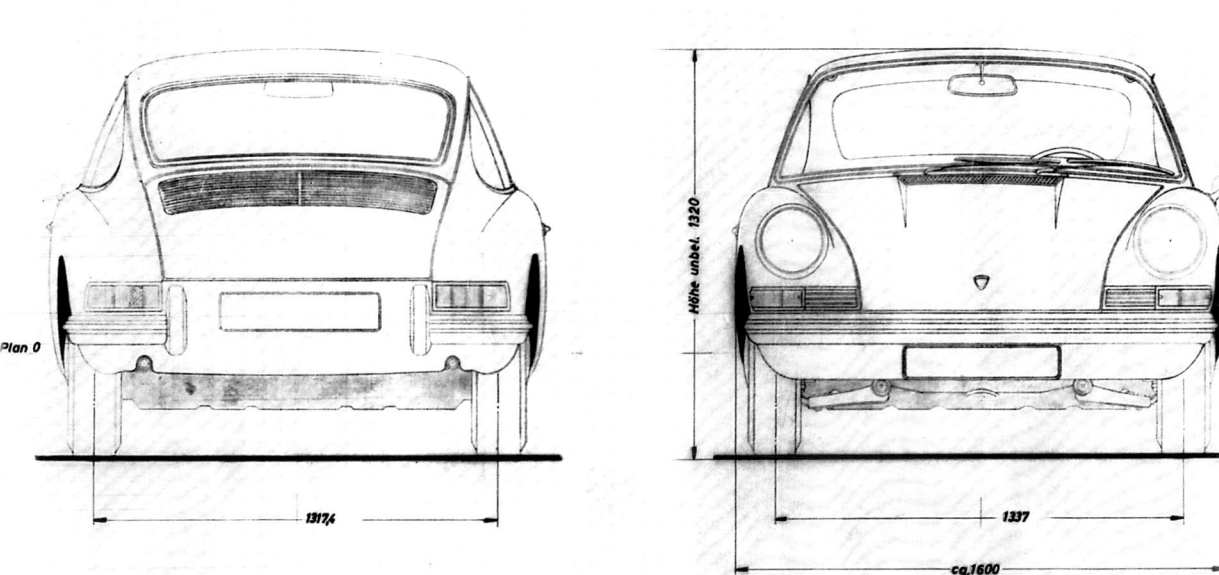

Zeichnung Typ 901 vom 24. April 1964
Drawing of the Type 901, April 24, 1964

LEBENSLAUF
PERSONAL CHRONICLE

FERDINAND ANTON ERNST »FERRY« PORSCHE

1909
Am 19. September wird Ferdinand Anton Ernst Porsche, genannt »Ferry«, in Wiener Neustadt (Österreich) geboren. Schulbesuch in Wiener Neustadt und in Stuttgart-Bad Cannstatt. Technische Ausbildung bei Bosch in Stuttgart sowie bei den österreichischen Steyr-Werken.

1931
Beginn der Tätigkeit als Konstrukteur im Konstruktionsbüro Dr. Ing. h.c. Ferdinand Porsche GmbH in Stuttgart.

FERDINAND ANTON ERNST ('FERRY') PORSCHE

1909
Ferdinand Anton Ernst Porsche, known as 'Ferry', was born in Wiener Neustadt (Austria) on September 19. He attended school in that town and later in Stuttgart-Bad Cannstatt. He received technical training at Bosch in Stuttgart and at the Austrian Steyr-Werke.

1931
Ferry Porsche began work as a designer in the Dr. Ing. h.c. Ferdinand Porsche GmbH design office in Stuttgart.

1932
Erweiterung der Aufgaben auf Versuchsüberwachung und Koordination, Mitwirkung an Konstruktion und Entwicklung des Auto Union-Grand Prix-Rennwagens.

1934
Leiter der Erprobungsfahrten der Volkswagen-Prototypen.

1935
Hochzeit mit der Stuttgarterin Dorothea Reitz (gestorben 1985). Aus der Ehe gehen vier Söhne hervor.

1938
Leiter der Porsche-Versuchsabteilung. Im selben Jahr erfolgt der Umzug des Konstruktionsbüros nach Stuttgart-Zuffenhausen.

1940
Übernahme der stellvertretenden Leitung des Gesamtbetriebes.

1932
His duties in his father's company were extended to include supervision and coordination of testing and cooperation on the design and development of the Auto Union Grand Prix racing car.

1934
Ferry Porsche managed the road testing program for the Volkswagen prototypes.

1935
Married Dorothea Reitz from Stuttgart (she died in 1985). They had four children, all sons.

1938
Managed the Porsche experimental department. This was the year in which the design office moved to Stuttgart-Zuffenhausen.

1940
Deputy general manager of the complete Porsche operation.

1945
Leiter der kriegsbedingt nach Gmünd in Kärnten (Österreich) verlagerten Firma.

1946
Ferry Porsche übernimmt im Juni die Gesamtverantwortung für das Unternehmen.

1948
Im Juni wird der 356 »Nr. 1« fertiggestellt, ein Mittelmotorsportwagen mit 35 PS.

1949
Nach dem Bau der ersten 52 Exemplare des Typs 356 in Gmünd, Rückkehr nach Stuttgart-Zuffenhausen. Wiederaufbau des Entwicklungsbüros unter der Leitung von Ferry Porsche und Vorbereitung der Serienfertigung.

1950
Beginn der Serienfertigung des Typ 356 in Stuttgart-Zuffenhausen.

1945
Managed the company when it moved to Gmünd, Carinthia (Austria) to avoid the risk of damage due to bombing raids.

1946
Ferry Porsche assumed overall responsibility for the company in June.

1948
The first 356 ('Number One'), a sports car with a 35-hp mid-engine, was completed in June.

1949
After the first 52 Type 356 cars had been built in Gmünd, the company returned to Stuttgart-Zuffenhausen. The development office was built up again under Ferry Porsche's management, and preparations were made for production of the 356.

1950
The Porsche Type 356 went into series production in Stuttgart-Zuffenhausen.

1959
Verleihung des großen Verdienstkreuzes der Bundesrepublik Deutschland durch Bundespräsident Theodor Heuss.

1965
Verleihung des Titels »Dr. techn. E.h.« durch die Technische Hochschule in Wien.

1972
Übernahme des Vorsitzes im Aufsichtsrat der in eine Aktiengesellschaft umgewandelten Dr. Ing. h.c. F. Porsche AG.

1975
Verleihung des Großen Goldenen Ehrenzeichens der Republik Österreich in Wien.

1978
Verleihung der Wilhelm-Exner-Medaille.

1979
Verleihung des Sterns zum Großen Verdienstkreuz der Bundesrepublik Deutschland anlässlich des

1959
Ferry Porsche received the Grand Cross of Merit of the Federal Republic of Germany from President Theodor Heuss.

1965
Awarded the honorary doctorate 'Dr. techn. E.h.' by the Vienna College of Advanced Technology.

1972
Became Chairman of the Supervisory Board of Dr. Ing. h.c. F. Porsche AG, which had been converted into a joint stock corporation.

1975
Presented with the Republic of Austria's Grand Medal of Honor in Gold in Vienna.

1978
Awarded the Wilhelm Exner medal.

1979
On the occasion of his 70th birthday, awarded the Star of the Grand Cross of Merit of the Federal Republic

70. Geburtstages durch den Ministerpräsidenten des Landes Baden-Württemberg, Lothar Späth.

1981
Verleihung der Goldmedaille der Société des Ingénieurs de l' Automobile.
Verleihung der Ehrenbürgerwürde der Stadt Zell am See (Österreich).

1984
Verleihung des Titels »Professor« durch Ministerpräsident Lothar Späth.

1985
Verleihung des Titel »Senator E.h.«, Universität Stuttgart.

1989
Verleihung der Wirtschaftsmedaille für herausragende Verdienste um die Wirtschaft Baden-Württembergs am 19. September durch den Wirtschaftsminister des Landes Baden-Württemberg, Martin Herzog.
Verleihung der Bürgermedaille der Stadt Stuttgart an-

of Germany; the presentation was carried out by Lothar Späth, Prime Minister of the State of Baden-Wuerttemberg.

1981
Awarded the Gold Medal of the Société des Ingénieurs de l' Automobile.
Granted the freedom of the town of Zell am See (Austria).

1984
Granted a professorship by Prime Minister Lothar Späth.

1985
Appointed Honorary Senator of Stuttgart University.

1989
On September 19, awarded the Economic Medal for outstanding services to the economy of the German State of Baden-Wuerttemberg by the Minister for Economic Affairs, Martin Herzog.
To celebrate his 80th birthday, presented with

lässlich seines 80. Geburtstages in Würdigung seiner großen Verdienste um die wirtschaftliche Entwicklung der Landeshauptstadt Stuttgart.

1990
Ehrenvorsitzender des Aufsichtsrates der Dr. Ing. h.c. F. Porsche AG, Stuttgart.

1994
Verleihung der Ehrenbürgerwürde der Stadt Wiener Neustadt am 21. September in Würdigung seiner besonderen Verdienste um die österreichische und niederösterreichische Wirtschaft.

1998
Ferry Porsche stirbt am 27. März in Zell am See.

the Stuttgart Citizens' Medal to acknowledge his meritorious service to the economic development of the State capital.

1990
Honorary Chairman of the Supervisory Board of Dr. Ing. h.c. F. Porsche AG, Stuttgart.

1994
On September 21, granted the freedom of the town of Wiener Neustadt in recognition of his exceptional services to the national and Lower Austrian economies.

1998
Ferry Porsche died on March 27 in Zell am See.

Dr. Ing. h.c. F. Porsche AG
Porscheplatz 1
70435 Stuttgart
Germany
www.porsche.com/museum

Inhalt & Abbildungen: Porsche AG, Historisches Archiv
Content & photos: Porsche AG, Historical Archives
Design & Layout: SEIDLDESIGN
Druck/Printing: GZD

Edition Porsche-Museum
1. Auflage 2009 | 1st edition 2009
ISBN 978-3-9812816-2-0

Alle Rechte vorbehalten. Die Rechte für die Verwendung der Abbildungen und Textbeiträge liegen bei der Dr. Ing. h.c. F. Porsche AG, Stuttgart. Ohne ausdrückliche Erlaubnis darf das Werk, auch Teile daraus, weder reproduziert, übertragen noch kopiert werden.

All rights reserved. Rights for use of illustrations and text are property of Dr. Ing. h.c. F. Porsche AG, Stuttgart. This work may not be reproduced, transmitted, or copied, in whole or in part, without express permission from the publisher.